Natural Food Preservatives

Natural Food Preservatives

edited by
Sourish Bhattacharya
Heba Hassan Abd-El Azim Salama

JENNY STANFORD
PUBLISHING

Published by

Jenny Stanford Publishing Pte. Ltd.
101 Thomson Road
#06-01, United Square
Singapore 307591

Email: editorial@jennystanford.com
Web: www.jennystanford.com

British Library Cataloguing-in-Publication Data
A catalogue record for this book is available from the British Library.

Natural Food Preservatives

ISBN 978-981-4968-64-5 (Hardcover)
ISBN 978-1-003-36776-5 (eBook)

Contents

Preface

Humans have been preserving food for generations, and both conventional and novel preservatives are now routinely used to guarantee the satisfactory preservation of food quality and safety. However, as eating patterns have changed and a focus on what is known as a "healthy diet" has gained popularity, there is also concern that reducing the use of natural preservatives could result in a loss of safety and protection from food poisoning. While certain preservatives face rising regulatory pressure, others—particularly those that are more natural—are getting more attention, importance, and acceptance.

Natural food preservatives are natural compounds, either natural or synthetic, that delay food breakdown brought on by oxidation, microbial development, and enzyme activity. However, in recent years, due to the assumed adverse effects of synthetic additives on health, consumers' preference of natural food preservatives is increasing day-by-day.

Chapter 1 introduces the microbiological spoilage of foods and beverages. It discusses various types of food spoilage caused by the microorganisms and their impact on food quality in various important food commodities. Chapter 2 discusses utilizing grape and winery by-product extracts as natural preservatives in sustainable food systems. Chapter 3 discusses utilization of nisin as a natural food preservative having the potential to replace chemical preservatives. Chapters 4 and 5 discuss major preservation technologies, including utilizing conventional and non-conventional preservatives. Further, Chapter 6 discusses the diverse content of preservatives for industrial food products. Chapters 7 and 8 discuss the basic chemistry and mode of action of natural food preservatives. Chapter 9 discusses plant-based food preservatives and Chapter 10 discusses the utilization of ε-polylysine as a natural food preservative.

Within the limitations of chemical preservatives, this book encourages the continued safe and efficient use of natural food preservatives. It provides comprehensive information on the actual use of main antimicrobial preservatives. It also combines the current and existing knowledge on the mode of action at the cellular physiology and biochemistry levels. We hope the book will help promote the logical development and application of natural preservatives in the future.

Sourish Bhattacharya
Heba Hassan Abd-El Azim Salama
Summer 2023

Chapter 1

Introduction to the Microbiological Spoilage of Foods and Beverages

Ashraf Mahdy Sharoba
Food Technology Department, Moshtohor Faculty of Agriculture, Banha University, Egypt
ashraf.sharouba@fagr.bu.edu.eg

Food spoilage is a complex ecological phenomenon, caused mainly by the biochemical activity of specific groups of microorganisms. Microbiological spoilage of foods and beverages is the main cause of quality deterioration in these commodities. Microorganisms (i.e., bacteria, molds, and yeasts) associated with spoilage as well as factors affecting their growth and determining their eventual dominance in foods and beverages are presented. During harvesting, processing, and handling operations, food may become contaminated with a wide range of microorganisms. Subsequently, during distribution and storage, only a small fraction of these will develop and cause serious deterioration. Which microorganisms will develop is dependent upon food derived or environmental factors. The loss of

Natural Food Preservatives
Edited by Sourish Bhattacharya and Heba Hassan Abd-El Azim Salama

ISBN 978-981-4968-64-5 (Hardcover), 978-1-003-36776-5 (eBook)
www.jennystanford.com

food quality for the most important food commodities depends on the microbiological load. Food spoilage may be caused by a wide range of reactions including some that are mainly physical or chemical and others due to the action of enzymes or microorganisms. The primary factors associated with food spoilage are associated with intrinsic food properties and (cross) contamination during harvesting, slaughter, and processing in combination with temperature abuse. For fresh foods, the primary quality changes may be bacterial growth and metabolism resulting in possible pH changes and formation of toxic compounds, off-odors, gas, and slime formation. Despite the fact that food spoilage is a huge economical problem, it is obvious that the mechanisms and interactions leading to food spoilage are very poorly understood. In this chapter, we will deal with the most important causes of microbial corruption in food and beverages and the most important and most famous types of microbial corruption in food and beverages and the extent of the risk of these types of microorganisms to the quality of food products and the health of consumers in addition to discussing how to reduce or prevent the occurrence of microbial corruption in food and beverages.

1.1 Introduction

1.1.1 Microbiological Spoilage of Foods and Beverages

Worldwide postharvest fruit and vegetable losses are as high as 30–40% and even much higher in some developing countries. Reducing postharvest losses is very important, ensuring that sufficient food, both in quantity and in quality, is available to every inhabitant on our planet. The prospects are also that the world population will grow from 5.7 billion inhabitants in 1995 to 8.3 billion in 2025. World production of vegetables amounted to 486 million tons, while that of fruits reached 392 million tons. Reduction of postharvest losses reduces the cost of production, trade, and distribution, thus lowering the price for the consumer and increasing the farmer's income.

Fresh fruits and vegetables are perishable and highly prone to these losses because they are composed of living tissues. These tissues must be kept alive and healthy throughout the process of marketing. These are composed of thousands of living cells that require care and maintenance. Therefore, the reduction of postharvest loss of fruit and vegetables is a complementary means for increasing production. It may not be necessary to considerably step up the production of fruits and vegetables with the growing demand if the postharvest loss is reduced to a great extent. The cost of preventing losses after harvest, in general, is less than preventing a similar additional amount of fruit and vegetable crops of the same quality. Attention to the concept of postharvest food loss reduction, as a significant means to increase food availability, was drawn by the World Food Conference held in Rome in 1974. The global dairy industry is impressive by large. In 2005, world milk production was estimated at 644 million tons, of which 541 million tons were cows' milk. The leading producers of milk were the European Union at 142 million tons, India at 88 million tons, the United States at 80 million tons (20.9 billion gallons), and Russia at 31 million tons. Cheese production amounted to 8.6 million tons in Western Europe and 4.8 million tons in the United States [1]. The vast array of products made from milk worldwide leads to an equally impressive array of spoilage microorganisms. A survey of dairy product consumption revealed that 6% of US consumers would eat more dairy products if they stay fresher longer [2]. Products range from those that are readily spoiled by microorganisms to those that are shelf-stable for many months, and the spoilage rate can be influenced by factors such as moisture content, pH, processing parameters, and temperature of storage. Microbiological determination of juices is considered one of the most important studies that could be taken before, during, and after the production of processed food. It is also used as an index, testing the sufficiency of the processing operation for the microbial aspects and the sanitary condition of the processing. The main purpose of microbial determination is the detection of pathogenic microorganisms to ensure that processed fruit is safe.

In 1998, FDA published the Guide to Minimize Microbial Food Safety Hazards for Fresh Fruits and Vegetables, recommending Good Agricultural Practices (GAPs) that growers, packers, and shippers implement to address the common microbiological hazards that may be associated with their operations [3]. These GAPs are organized into the following eight categories:

I. Water
II. Manure and municipal biosolids
III. Worker health and hygiene
IV. Sanitary facilities
V. Field sanitation
VI. Packing facilities sanitation
VII. Transportation
VIII. Traceback

[4] reported that some low acids canned food products have viable spore-forming organisms (*Clostridium* and *Bacillus* group) that could have survived heat processing. Also, canned foods belonging to the high acid foods including fruit juices like papaya were studied and viable yeast and non-spore-forming organisms were found in processed papaya juice. Their presence, however, was the result of can leakage. [5] reported that processes involved in the manufacture of fruit juice significantly affect the composition, moisture, and solids content and influence the type of spoilage organisms present in the juices. A high moisture content favors the growth of bacteria and yeast.

The reduction of solids by processing may change the oxidation-reduction potential and the presence of residual sugars may result in alcoholic fermentation for canned food products, especially at room temperature. Organisms are very heat resistant and can withstand a processing temperature of 100°C for 90 min; the organism may produce off-flavor and flat sour type of spoilage in the juice. Also, a few yeasts and molds are thermo-acidic and could cause many troubles in the juice industry, such as off-flavor, gases, fermentation, and bad accepted odor in the product.

The general method for reducing a microbial contamination level of fresh vegetable juice is a washing practice of raw materials in water with a contamination of ozone and/or chlorine treatment. However, it was a limited effect on microflora and sometimes can contaminate the products [6]. Microbial growth in fruit and fruit products is of serious concern, as it causes decay, loss of nutritive and organoleptic properties, and production of toxic substances [7]. Owing to the low pH of most fruit products, fungi are often the predominant microorganisms while only a few bacteria are sufficiently aciduric to be important [8]. The pH of fruit juices and beverages had previously been thought to be too low to support the growth of spore-forming bacteria [9]. [10] found that coliform counts in these samples (fresh mangoes, pure, and juices) ranged between 1.4 × 103 and 5.4 × 104 cfu/g. Pasteurization reduced the coliform count of raw pure to < 5 cfu/g. Yeast and molds are considered the primary spoilage microorganisms in foods and beverages with high levels of sugars and low pH.

Most fruit juices are acidic enough and have sufficient sugar to favor the growth of yeast [11]. Molds are generally considered to be the least important group of microorganisms causing spoilage in fruit juice because of their limitation and inability to grow in the absence of air [12], with the exception of a few molds such as *Penicillium* and *Aspergillus* which are spore forming species [13].

The most resistant spoilage yeast isolated from fermented pasteurized fruit juices and nectar samples would be normally destroyed during vacuum operation at the industrial level. The yeast is isolated from a commercial beverage after the pasteurization process. The unpasteurized fresh juices can support the growth of both spoilage and pathogenic bacteria because they have high water activity (aw) and enough nutrients for microbial growth. If a bacterial pathogen is present on or in the product or on any processing surface, it will have an almost unlimited food source for growth in the unpasteurized juice, resulting in a foodborne outbreak [14]. In processed fruit juice, bacteria are the most diversified microorganisms causing its spoilage. Lactic acid bacteria *Acetobacter* and *Acetomonas* found on fruit surfaces comprise the most frequent spoilers of fruit juice because they exist on the surface of plants and fruits growing at the expense of secreted plant materials [15]. *Enterobacter* spp. are commonly found on visually all types of unsound fruits in the wash water. The presence of coliform mostly of the *E. aerogenes* type in both fresh and frozen fruit juices has been attributed to their being natural flora of fruits which may be introduced into the fruit juice if improperly processed [15].

[16] studied the quality of some market fruit drinks. Mango and guava drinks sold in India were evaluated for quality in initial microbial load and microbial load after six months of storage at room temperature (26–37°C). Analysis indicated that drinks with 14–16% total soluble solids and 0.21–0.24% acidity were determined stable during storage, and initial aerobic plate count showed the presence

of mold, yeast, and bacteria. After six months of storage, the microbial load had declined in all drinks. Isolated microorganisms were nonpathogenic.

The microbiological quality of juice fruits and nectar is dependent on a number of factors such as raw materials and sanitation during the process. Juices and nectar had a total bacterial count of less than 100 cfu/g. The low numbers of bacterial count may be related to the effect of the processing and storing method of the juice, yeast and molds and coliforms which were not detected in nectar samples (Pa3 : or1 and Pa3 and Ap1). The results may be in agreement with the previous results found by [17] for mango puree. [18] stated that during thermal processing of fruit products, molds are usually killed, but ascospores of heat-resistant molds lead to spoilage of processed fruit products.

Fungi are the major causes of spoilage of fruit and vegetables. The presence of aflatoxin and other mycotoxin-producing mold contamination on fruit surfaces has been reported to generally end in the juice [19]. Shelf-stable juice and beverages, processing in the range of 85–95°C, are targeted for inactivating pathogenic foodborne microorganisms. These thermal processes are inefficient and are thermo-tolerant leading to spoilage [20]. [21] reported that the concentration methods and additives reduced the total viable bacterial count (TVBC) and mold and yeast counts of concentrated apricot and peach juices.

Fruit juice can be spoiled due to the growth of microorganisms. Yeasts and molds, *Lactobacillus*, *Leuconostoc*, and thermophilic *Bacillus* are common spoilage microorganisms of orange juice [22]. *Escherichia coli*, *Saccharomyces cerevisiae*, and *Listeria innocua* are associated with apple juice or apple cider spoilage [23].

Many organisms, especially acid-tolerant bacteria and fungi (yeasts and molds), can cause spoilage in the fruit juice using fruit as a substrate. Discoloration, off-flavors, and off-odors may develop in the product. The most common fruit juice spoilers are yeasts [24]. Low pH values of the juices prevent the survival of many types of bacteria and provide a suitable environment for yeasts, molds, and a few groups of aciduric bacteria. It was demonstrated that *B. coagulans* is responsible for the flat sour type spoilage in acidic beverages. It has the ability to grow at pH values as low as pH 4.0 [25].

Even though the fruit juices have low pH values and thermal pasteurization can damage spoilage microorganisms, a bacterium called *Alicyclobacillus acidoterrestris* is found to survive under normal pasteurization conditions. It is a nonpathogenic, spore-forming, and thermoacidophilic bacterium that is isolated from several commercial pasteurized orange and apple juices [20]. It can grow at a pH range of 2.5–6.0 at temperatures 25–60°C [26]. The time needed to destroy 90% of this bacterium at 90°C was found to be 10.0–20.6 min. This high D-value makes the pasteurization process unsuccessful to destroy the microorganism in orange concentrates [27]. In recent times, Alicyclobacilli are considered quality control microorganisms in the manufacturing of acidic beverages [28]. This microorganism causes economic losses and off-flavors in fruit juices [25].

Fresh fruit and vegetable juices are highly contaminated by microorganisms. The microbial flora in the vegetable juice is thus likely to be similar to the flora in raw vegetables. Many microorganisms, in particular acid-tolerant bacteria and fungi (yeasts and molds), can use fruit as substrate and cause spoilage, producing off-flavors and odors, discoloration of the product, and if the contaminating microorganisms are pathogens, they could also cause human illness [24].

Although many yeast species can tolerate a broad range of pH, from 1.5 to 10.0, most of them prefer proliferation in a slightly acidic environment ranging from pH 3.5 to 6.0. More than 800 species of yeasts have been identified. However, a small portion of these are considered spoilage ones [29]. Yeast species commonly isolated from fruit juices are *Rhodotorula rubra*, *Candida lambica*, *C. sake*, *Kloeckera apis*, *Hanseniaspora uvarum*, *S. cerevisiae*, *Zygosaccharomyces bailii*, *Z. rouxii*, *Torulaspora delbrueckii*, and *C. intermedia* [24, 29–32]. *H. uvarum* is demonstrated to be the only species that was isolated from both pasteurized and unpasteurized orange juices. *T. delbrueckii*, *C. intermedia*, and *S. cerevisiae* are fermentative yeast species and they are not able to survive after processing of the product [30]. In [33], using standard microbiological procedures, 146 fruit juice samples sold in Tripoli Libya were examined. Total bacterial counts (TBC) of samples examined ranged between <1 and 3 ×105 cfu/ml (mean = 5 × 104). Almond juice showed the highest TBC ranging between 1.8 × 103 and 3 × 105 (mean = 1.7 × 105). *Staphylococcus aureus* was detected in 8 (5.5%) samples, *Streptococcus* spp. in 4 (2.7%), coliforms in 33 (22.6%), *E. coli* (none were of serogroup 0157) in 3 (2.1%), *Klebsiella pneumonia* in 17 (11.6%), *Aeromonas* spp. in 3 (2.1%), *Pseudomonas aeruginosa* in 6 (4.1%), *C. albicans* in 18 (12.3%), *Candida* spp. in 109 (74.7%), and other yeasts in 85 (58.2%).

Thirty-eight fruit salad samples, including cantaloupe, citrus fruits, honeydew, pineapple, cut strawberries, and mixed fruit salads, and 65 pasteurized fruit juice samples (apple, carrot, grapefruit, grape and orange juices, apple cider, and soy milk) were purchased from local supermarkets in the Washington, DC area and tested for fungal contamination. The majority of fruit salad samples (97%) were contaminated with yeasts at levels ranging from <2.0 to 9.72 log10 of colony-forming units per gram (cfu/g). Frequently encountered yeasts were *Pichia* spp., *C. pulcherrima*, *C. lambica*, *C. sake*, *Rhodotorula* spp., and *Debaryomyces polymorphus*. A low number of *Penicillium* spp. were found in pineapple salads, whereas *Cladosporium* spp. was present in mixed fruit and cut strawberry salads. Twenty-two percent of the fruit juice samples tested showed fungal contamination. Yeasts were the predominant contaminants ranging from <1.0 to 6.83 log10 cfu/ml. Yeasts commonly found in fruit juices were *C. lambica*, *C. sake*, and *R. rubra*. *Geotrichum* spp. and a low number of *Penicillium* and *Fusarium* spp. (1.70 and 1.60 log10 cfu/ml, respectively) were present in grapefruit juice [24]. Therefore, vegetable juice should undergo some type of processing to inactivate most of microorganisms. The main requirement that a process must meet is to ensure the microbial safety of the product while preserving the sensory and nutritional characteristics to obtain products similar to fresh vegetable juice [34]. The microbiological safety of different fruit juices being served in cafes/restaurants in Jimma town was assessed based on the isolates API-20E (Analytical Profile Index) used to test for 20 substrates to differentiate Enterobacteriaceae profile from February 2005 to July 2006. A total of 90 samples (30 samples each for avocado, papaya, and pineapple), collected from six randomly selected cafes and/or restaurants in Jimma town, were analyzed. The results of mean aerobic mesophilic bacteria counts (cfu/ml) of avocado, papaya, and pineapples were 8.0×10^6, 3.1×10^7, and 7.9×10^6, respectively. The counts of yeasts were relatively higher in avocado (4.5×10^5 cfu/

ml) and pineapple (5.0×10^6 cfu/ml) as compared to that of papaya (6.2×10^3 cfu/ml) [35]. Microbiological and physicochemical qualities of tomato (*Lycopersicon esculentum*) puree production line (ripe tomato, washing, cutting, pounding, bleaching, straining, bottling, and pasteurization) and its preservation in Togo, West Africa, were studied using the Hazard Analysis and Critical Control Point (HACCP) method [36]. Samples generated during the steps described previously were analyzed by determining sensory, chemical, and microbiological characteristics. Samples were analyzed using MPN for coliform populations and plate count methodology for other bacteria. The microorganisms involved in the spoilage of the opened products were molds of genera *Penicillium*, *Aspergillus*, *Fusarium*, *Geotrichum*, *Mucor*, and Gram-positive *Bacillus* bacteria. The preserved tomato puree exhibited a pH value of 4.3, 90% water content, 0.98 water activity (aw), and an average ascorbic acid level of 27.3 mg/100 g. [37] studied the quality assessment of industrially processed fruit juices available in Dhaka city, Bangladesh. He reported that the microbial qualities of all products were within the limits of the Gulf standards (the recommended microbial standards for any fruit juice sold in the Gulf Region). Based on the above results, it is suggested that processed juices be prepared under hygienic conditions. [38] studied an assessment of the microbiological quality of some commercially packed and fresh fruit juice available in Dhaka City. He reported that, in general, microbial growth in fruit juice is restricted by using preservatives and also through other environmental factors during production. Still, the presence of some indicator organisms in fruit juice is responsible for causing foodborne diseases and even death. So, from a public health point of view, it is quite important to know the microbiological quality of the fruit juices available in the market. From all the assessments, it was determined that the microbial quality of commercially packed juice was fairer than that of fresh juice collected from the local market. [39] studied the microbiological and chemical quality assessment of some commercially packed fruit juices solid in Nigeria. He reported that samples of orange, lemon, pineapple, punch, and guava juice contained total aerobic mesophilic bacterial counts ranging from 1.4×10^4 to 2.6×10^5 cfu/ml and the fungal count ranges from 1.4×10^3 to 1.7×10^5 cfu/ml, while the coliform counts range from 1.1×10^4 to 6×10^4 cfu/ml. The total microbial counts are within acceptable standards for human consumption. The level of bacterial counts and fungal count and the absence of heavy metals conform to the standard specification of the National Agency for Food and Drug Administration and Control (NAFDAC) and Standard Organization of Nigeria (SON). The presence of coliforms in some of the foods which are maintaining the standard specifications of the NAFDAC and SON. The presence of coliforms in some of the fruit juice calls for strict adherence to GMP and effective HACCP applications.

1.2 Microbiology Quality of Juices and Nectars

Microbial growth in fresh juices and juice nectars is of serious concern, as it causes decay, loss of nutritive and organoleptic properties, and production of toxic substances. Owing to the low pH of most fruit and vegetable products, fungi are often the predominant microorganisms while only a few bacteria are sufficiently aciduric to be important [40]. The pH of juices and beverages had previously been thought to be too low to support the growth of spore-forming bacteria [9]. The microbiological quality of juice fruits and nectar is dependent on a number of factors such as raw materials and sanitation during the process. Juices and nectar had a total bacterial count of less than 100 cfu/g. The results may be in agreement with the previous results found by [17] for mango puree.

The data in Table 1.1 revealed that the total viable bacterial count was 53 and 60 cfu/g for papaya juice and papaya nectar, respectively. Also, Table 1.1 shows the evaluation of the total microbial counts, yeasts, molds, and coliform. It could be observed that the total bacterial count as well as yeasts and molds counts were highest with increase in orange juices, carrot juices, apricot nectar, and carrot nectar samples. Coliform bacteria count was less than 30 in most juices and nectar this result may be in agreement with [41]. From Table 1.1, it is clear that many viable yeast and mold and bacteria were detected in all fresh juices and nectars. The total viable bacterial count of yeast and mold increased in fresh nectar samples than in fresh juices samples which may be due to the addition of water to nectar samples.

Moreover, the increase in the total bacterial count and yeast and mold in fresh juice and nectar samples may be due to the contamination during manufacture.

Table 1.1 Microbiology quality of fresh juices and nectars (cfu/g)

Juices	TVBC	Y & M	Coliform
Papaya	53	45	<30
Mango	56	42	ND
Orange	71	47	<30
Apricot	68	44	ND
Carrot	79	53	<30
Nectars			
Papaya	60	47	<30
Mango	64	45	ND
Orange	73	44	ND
Apricot	79	56	<30
Carrot	82	58	<30

TVBC, total viable bacterial count; Y & M, yeasts and molds.

1.3 Beneficial Microbes in Food

The changes in the present-day consumers' lifestyle have led to a vital change in the marketing trends of the food sector. It was mainly the advances in understanding the relationship between nutrition and health that resulted in the development of the concept of functional foods, which means a practical and new approach to achieving optimal health status by promoting the state of well-being and possibly reducing the risk of disease. International recognition of functional foods has resulted in the recent development of this field, which resulted in significant growth in the fruit juices market, which has attracted the attention of fruit growers, fruit juice distributors, and processors to meet the demand for fruit juices. Fruit juice beverages with nutraceuticals are now gaining more importance in the fruit juice market [42]. The consumption of foods and beverages containing functional prebiotics and probiotics is the current global consumer trend.

During the last 2–3 decades, attempts for improving the human health status are focusing on ways to modulate the indigenous intestinal flora by live microbial adjuncts, now called "probiotics".

Comprising ca. 65% of the functional food world market, probiotics represent the major and still growing segment of this huge market, estimated to exceed a total volume of US$75 billion. The most typical active components of probiotic products are lactic acid bacteria, including Bifidobacteria, Lactobacilli, and Enterococci. Health claims related to probiotics are numerous, but include maintenance of normal/healthy intestinal flora and protection against infections, alleviation of lactose intolerance, and stimulation of the immune system. Strains with proven beneficial effects may be consumed in relatively high numbers in products and are collectively called probiotics. Besides, the probiotic products, there are also prebiotic functional foods, in which the development has increased lately. A prebiotic is a specific ingredient that allows changes, both in the composition and/or activity of the gastrointestinal microflora that confers benefits upon host well-being and health [43].

In general, the suppliers of functional ingredients try to prove the efficacy of a specific substance and sell it to a wide range of food industry companies, thereby creating specific formulations. In this sense, innovative suppliers of food ingredients are of high relevance, in particular for product innovations of SME food companies. The world demand for prebiotics is estimated to be around 167,000 tons and 390 million euros. Among them, FOS, inulin, isomalto-oligosaccharides (IMOs), polydextrose, lactulose, and resistant starch are considered the main prebiotic components. Oligosaccharides, such as soy oligosaccharides (SOSs), galactooligosaccharides (GOSs), and xylooligosaccharides (XOSs), are also marketed in Japan. Inulin and oligofructose, nondigestible fermentable fructans, are among the most studied and well-established prebiotics.

Commercial probiotic products include primarily dairy products (fermented milk, yoghurt, ice cream, cheeses) and to a lesser extent nondairy products (beverages, breakfast cereals, fermented meats, dry foods) and dietary supplements. Novel products containing probiotics have been launched recently, mainly beverages based on fruits and cereals. However, fruit juices are more adverse environments than fermented milks for probiotics as they have high acidity and high levels of phenolic compounds. Consumers, on the other hand, demand that the product they purchase contains the concentration of probiotic cells stated on the package at the time of consumption. Therefore, identifying the factors influencing probiotic survival in juices and developing ways to enhance probiotic survival during storage is an important area of research with considerable impact on the food industry [44].

1.4 Probiotic Microorganisms

Probiotics primarily belong to the genera *Lactobacillus* and *Bifidobacterium*; however, other microorganisms, including propionibacteria, leuconostoc, pediococci, enterococci, and *E. coli*, have also been considered probiotic cultures. A summary of potential probiotic species is provided in Table 1.2. Probiotic preparations may include one or more different microbial strains or species.

Table 1.2 Microorganisms that have been identified as probiotic microflora

Microflora	Species
Lactobacilli	*L. acidophilus*
	L. rhamnosus
	L. reuteri
	L. casei
	L. gasseri
	L. plantarum
	L. johnsonii
Bifidobacteria	*B. bifidum*
	B. longum
	B. breve
	B. adolescentis
	B. infantis
Enterococci	*Enterococcus faecium*
	E. faecalis
Yeast	*S. boulardii*
Lactococci	*Lactococcus lactis subsp. lactis*

Source: [45]

1.4.1 Health Effects of Probiotics

The ability of probiotic microorganisms to survive and multiply in the host depends on their colonization in the gut. The probiotic bacteria should be metabolically stable and active in the product, survive and pass through the upper digestive tract in large numbers, and have beneficial effects when in the intestine of the host. The health benefits of probiotic bacteria, such as *Lactobacillus* spp. and *Bifidobacterium* spp., have led to numerous scientific studies being performed on their therapeutic impact on various infections and diseases in humans.

1.4.2 Improvement of Lactose Digestion

Lactose intolerance is a problem for ≥70% of the world population who have a low amount of intestinal beta-galactosidase activity and for whom lactose behaves like an osmotic nondigestible carbohydrate. Probiotics have been shown to improve lactose digestion by reducing intolerance symptoms as well as by slowing transit. Clinical research on probiotics and lactose intolerance suggests that an increased number of probiotics in the intestinal tract might contribute to the reduction of lactose intolerance symptoms. It is thought that the release of lactase by broken Lactobacilli in the small intestine and fermentation of lactose by probiotic bacteria in the colon help reduce the symptoms of lactose intolerance. The use of both families of probiotics may be helpful for people with lactose intolerance. Probiotic strains behave differently, and there is some evidence that suggests strains, concentrations, and preparations are likely effective against lactose intolerance.

1.4.3 Hypercholesterolemic Effect

L. acidophilus NCFM and *L. sporogenes* were able to assimilate cholesterol during growth and decrease the concentration of cholesterol in the growth medium. Uptake of cholesterol occurred only when the culture was grown anaerobically in the presence of bile, such as would be expected to occur in the human gastrointestinal tract. This may have important implications in preventing the reabsorption of cholesterol back into the systemic circulation. The administration of *L. plantarum* leads to a reduction in cardiovascular disease risk factors and could be useful as a protective agent in the primary prevention of atherosclerosis in smokers. Also, several studies have suggested that ingestion of probiotic microorganisms may lead to lower serum cholesterol levels and can sequent to a reduced risk of arteriosclerosis and other coronary heart diseases. The ability of probiotics to lower cholesterol levels is due to their ability to produce bile salt hydrolase (BSH). Bile salt hydrolase is an enzyme that converts bile into deconjugated bile salts. Deconjugated bile salts cannot be reabsorbed by the body; thus, more bile acid is lost in the feces and less cholesterol is reabsorbed and sent to the liver. Since about 95% of bile salts are normally reabsorbed, this enzyme can have a great effect on cholesterol levels. The production of bile salt hydrolase by probiotics appears to lower cholesterol by reducing the amount of cholesterol absorbed in the gut and increasing the amount of cholesterol pulled from the blood by the liver to make bile. A study done by *L. plantarum* PH04 (dose 4 × 108 cfu/ml) confirmed the safe use of this strain as a probiotic, and the cholesterol and triglyceride level in blood was significantly reduced. This was assumed due to decreased intestinal absorption of lipids or increased lipid catabolism. *L. casei* F0822 is a promising probiotic strain as a functional component for decreasing serum cholesterol levels. [46] reported that *L. plantarum* EM showed high cholesterol removal by growing, resting, and even dead cells based on the high cholesterol-binding capacity of its cell wall fraction. Therefore, *L. plantarum* EM could be a potent probiotic to reduce serum cholesterol regardless of its viability.

1.4.4 Stimulation of Immune System

The intestine is the body's largest immune organ in the host. The intestinal microflora has the metabolic activity of the intestine and is equivalent to that of the liver. Probiotic bacteria may directly enhance the function of the immune systems by increasing the number of immunoglobulin (IgA) plasma cells, improving phagocytosis, and increasing the proportion of T lymphocytes and natural killer cells. Changing the composition or activity of the intestinal microflora indirectly influences the body's immune function. Lactic acid bacteria as probiotics are an attractive, safe way to regulate and enhance the secretory response and particularly IgA production could be helpful in protecting against infection. The immune system acts to protect the host from infectious agents in the environment (bacteria, fungi, viruses, and parasites) and from other noxious insults. Increasing the intake of some nutrients containing probiotic bacteria could improve some aspects of the host immune function. Some positive of probiotics like protection of the host from intestinal infections. They mentioned that the resistance of people to disease could be improved upon by promoting the consumption of fermented foods containing probiotic bacteria because they are a good immunostimulatory agent. [47] studied the effect of *L. johnsonii* La1 on immune function and serum albumin in malnourished aged mice. They concluded that the intake of *L. johnsonii* La1 enhanced the intestinal (IgA) production and helped recover nutritional status and systemic immune responses in aged mice. It is possible that *L. johnsonii* La1 may contribute to immune system recovery in immunocompromised hosts such as elderly humans. [48] reported about the capacity of probiotics to improve the immune system. They found that the combination of nutrients and probiotic bacteria of yogurt best performed this role, appearing to be more efficient in attenuating the immune depression caused by exhaustive exercise than the whey probiotic beverage. This improvement in the immune system could be related to the improvement in the plasma lipid profile caused by the probiotic supplementation and also to the antioxidant power of probiotic supplementation.

1.4.5 Anticarcinogenic Action

There is some preliminary evidence that probiotic microorganisms can prevent and delay the onset of certain cancers. This stems from the knowledge that members of the gut microflora can produce carcinogens such as nitrosamines. Therefore, administration of Lactobacilli and Bifidobacteria could theoretically modify the flora, leading to decreased β-glucuronidase and carcinogenic levels, The probiotic may reduce the risk of cancer by several mechanisms, which include reducing enzymes associated with the conversion of procarcinogens to carcinogens, stimulating the host immunological defense, and altering the pH of the colon rendering the environment less conductive to the development of cancer. An in vitro study mentioned that *L. rhamnosus* GG and *Bifidobacterium* ssp. showed a decrease in the viability of carcinogenic aflatoxin in the lumen [49]. Probiotic Lactobacilli and Bifidobacteria strains decreased the quantity of such microbial enzymes as β-glucuronidase, β-glucosidase, nitroreductase, and urease, which were involved in the metabolic activation of miscellaneous mutagens and carcinogens. Furthermore, the regular consumption of some probiotics may decrease the levels of enzyme mutagens. They indicated that probiotics have anticarcinogenic effects and reduce the risk of colon and intestinal cancer. Probiotics may suppress the growth of bacteria that convert procarcinogens into carcinogens, thereby reducing the amount of carcinogens in the intestine. The anticancer activity through induction of apoptosis of cancer cells seems to be a promising approach for use of some probiotic strains as support therapy or disease prevention. [50] mentioned that anticancer assessments show high cytotoxicity for protein nature-secreted metabolites of *L. lactis* subsp. *lactis* 44Lac and *L. plantarum* 15HN on different human cancer cell lines.

1.5 Improvement of Rotavirus and *C. difficile* Diarrhea

[51] found that the addition of *L. acidophilus* to oral rehydration therapy was effective in the treatment of children with acute rotavirus diarrhea by decreasing the duration of rotavirus diarrhea. *L. monocytogenes*, *Salmonella*, and *Helicobacter pylori* growth is inhibited by producing antimicrobial substances such as organic acids, bacteriocins, hydrogen peroxide, and so on. *L. plantarum* and *L. rhamnosus* are helpful in reducing antibiotic associated diarrhea caused by *Clostridium difficile*. Probiotic *L. rhamnosus* GG, *B. lactis* Bb-12, and *B. animalis* Bb-12 are good for shortening the duration of rotaviral infection in children. At the same time, *L. rhamnosus* GG improves the rotavirus-specific IgA production. The symptoms of irritable bowel syndrome have been improved when probiotics are used as a supplement with the standard therapy. [52] showed that the antimicrobial activities of synbiotic carrot juice could differ in their antagonistic activities against a diarrheal causing organism, which could be due to the metabolite secreted by the lactic acid

bacteriocin, a special type of organic acid. Diarrhea is a major world health problem, responsible for several million deaths each year. While the majority of deaths occur amongst children in developing countries, it is estimated that up to 30% of the population even in developed countries are affected by foodborne diarrhea each year. Probiotics can potentially provide an important means to reduce these problems. The ability of probiotics to decrease the incidence or duration of certain diarrheal illnesses is perhaps the most substantiated health effect of probiotics. Probiotics can prevent or ameliorate diarrhea through their effects on the immune system.

1.6 Thermal Process of Fruit Juices

Fruit and vegetable juices and beverages are generally preserved by thermal processing, currently being the most cost-effective means of ensuring microbial safety and enzyme deactivation. However, thermal treatments may induce several chemical and physical changes that impair the organoleptic properties and may reduce the content or bioavailability of some nutrients; in most cases, these effects are strongly dependent on the food matrix. Moreover, the efficacy of treatments can also be affected by the complexity of the product and microorganisms.

1.7 Thermal Destruction of Microorganisms

In the thermal killing of unicellular organisms, the surviving population decreases logarithmically (log10) as the heating time increases linearly. This first-order relationship is quantified by *D* and *z* values. The *D* value is the amount of time, expressed in minutes, at a given temperature required to reduce a population by 90%. The *z* value is the amount of change in temperature, expressed as °C, that will shift the *D* value in either direction by 90%. For example, if a microorganism heated at 110°C has a *D*110°C value of 10 min and a *z* value of 10°C, its *D*120°C value would be 1.0 min if the heating occurred at 120°C, and the *D*100°C value would be 100 min if it occurred at 100°C. These particular *D* and *z* values are typical for many bacterial spores. Some bacteria have evolved mechanisms to cope with adverse environmental changes. *B. subtilis* produces a set of stress proteins that can be induced by exposure to mild heat or mild salt concentrations. Induction of these stress proteins enables *B. subtilis* to survive exposure to higher temperatures and salt concentrations.

Many approaches alternative to thermal treatments have been tested and successfully proposed for juices, but thermal processing still remains the most cost-effective tool to ensure microbial safety and enzyme deactivation. Some drawbacks of thermal processes are the slow conduction and convection heat transfer, and the negative effect of overprocessing on the sensory, nutritional, and functional properties. In most cases, these effects are strongly dependent on the food matrix. Moreover, the efficacy of thermal treatments can also be affected by the complexity of the product and microorganisms.

For example, some results of heat treatments for some juices, [53] reported the total activity of three important enzymes and their distribution in mango pulp and peel which were expressed as units/10 g fresh fruits. The results were 4.5 and 39.0, 0.55 and 0.7, and 3.5 × 103 and 180.0 × 103 for the pectinesterase, polygalacturonase and polyphenolase, respectively. Also, the results showed that the enzymes were higher in mango peel than in pulp and that was very important in technology. The same author calculated the *D* and *Z* values for mango (PE) and polyphenol oxidase (PPO); *D* values ranged from 20 to 15 min at 65–85°C and 33.5–10 min at 65–85°C for PE and P.P.O., respectively. While *Z* values were 18.5°C for PE and 26.25°C for P.P.O. From these results, it is obvious that the *Z* value of mango P.P.O. was higher than that of P.E. of the same origin. [54] estimated the thermal stability of three forms of pectinesterase (PMEI, PME2, and PME3) purified from tomatoes, which were evaluated at pH 4.5. The time-dependent, heat-inactivation curve for the forms exhibited exponential behavior in the temperature range 70–90°C. Determination of *D* and *Z* parameters revealed different heat-inactivation kinetics among the three isoenzymes. PME1 and PME3 had similar Z value, that is, 24°C, whereas PME2 had a value of 15°C. [55] studied the effect of canning processing steps on pectin methyl esterase (PME) activity of mango and guava nectars and found that PME activity was 87.2 and 135.8 (PME Ux106/g) for raw mango and guava nectar, respectively. Thermal processing at 90°C/35 min, 95°C/25 min, and 100°C/15 min reduced the PME activity to 6.3 and 6.6, 10 and 26.6, and 35.0 and 40.0 (PME U/106/g) for mango and guava nectar, respectively. Also, he determined the activity of PME of orange nectar and found that PME activity was 198.0 and 66.7 for raw and after preheating orange nectar, respectively. Thermal processing at 90.2°C/30 min, 95.2°C/20 min, and 100°C/15 min reduced the activity to 0.3, 4.5, and 16.5 units (PME Ux106/g for these thermal processes, respectively.

References

1. Kutzemeier, T. (2006). 27th World Dairy Congress in Shanghai, China. European Dairy Magazine, 7: 34–36.
2. Lempert, P. (2004). Waste not, want not. Progressive Grocer, 83: 18.
3. FDA (1998). Food and Drug Administation, Guide to Minimize Microbial Food Safety Hazards for Fresh Fruits and Vegetables. http:///www.cfsan.fda.gov/~dms/prodguid.html, viewed February 4, 2008.
4. Acevedo, T.P; Rillo, B.O; Sales, A.C; Tapang, N.; and Siongco, I.M. (1982). Spoilage organisms of thermally processed foods. NRCP Res. Bull., 37(1): 107–142.
5. Pierson, M.D. and Corlett, D.A. Jr. (1992). HACCP: Principles and Applications. Van Nostrand Reinhold, N.Y. pp. 212.
6. Nguyen-the, C. and Carlin, F. (1994). The microbiology of minimally processed fresh fruit and vegetable. Crit Rev Food Sci. Nutr., 34: 371–401.
7. Truong-Meyer, X.M.; Strehaino, P.; and Riba, J.P. (1994). Thermal treatment of yeast cells in a strawberry product- enumeration procedure. Microbial Biotechnol., 41: 262–266.
8. Splitstoesser, D.F. (1996). Microbilogy of fruit products. In L.p. somogyi, H.S.Ramaswamy, and Y.H. Hui (Eds.), Processing Fruit: Science and Technology (pp. 261–292). Basel: Technomic.
9. Blocher, J.C. and Busta, F.F. (1993). Bacterial spore resistance to acid. Food Technol., 37: 87–99.
10. O'Connor-Shaw, R.E.; Guthrie, J.A.; Dunlop, K.J.; and Roberts, R. (1995). Coliform in processed mango: Significance and control. International Journal of Food Microbiology, 25: 51–61.
11. Hooves, D.G. (1997). Manually processed fruit and vegetable: Reducing microbial load by non-thermal physical treatment. Food Technol. J., 51(6): 66–69.

12. Parish, M.E. (1991). Microbial concern in citrus juice processing. J. Food Technol., 45(4): 128–132.

13. Parish, M.E. and Higgin, D.P. (1989). Yeast and mould isolated from spoiling citrus products. J. Food Production, 52: 261–263.

14. Cook, K.A.; Dobbs, T.F.; Haldy, W.G.; Wells, J.G.; Barrett, T.J.; Puhr, N.D.; Lancette, G.A.; Bodager, D.W.; Toth, B.L.; Genese, C.A.; Highsmiith, A.K.; Pilot, K.E.; Finelli, L.; and Swerdlow, D.L. (1998). Outbreak of Salmonella serotype Hartford infection associated with un-pasteurized orange juice. JAMA, 280(17): 1504–1509.

15. Frazier, W.C. and Westhoff, D.C. (1998). Food Microbiology, 4th ed (pp. 196–215). New Delhi: McGraw Hill.

16. Kalra, S.K; Tandon, D.K; and Sengh, B.P. (1991). Quality evaluation of some market fruit drinks. Cent. Inst of Hort. For N. Plains. Lucknow 226019, India – Indian food paker 45(3): 48–53.

17. Hesham, M.H.S. (1998): Preservation of mango and mango pulp by radiation. Ph.D. Thesis, Fac. of Agric., Suez Canal Univ., Egypt.

18. Rajashekhara, E.; Suresh, E.R.; and Ethiraj, S. (1998). Thermal death of ascospores of neosartorya fischeri ATCC 200957 in the presence of organic acids and preservative in fruit juices. J. Food Protection, 61(10): 1358–1362.

19. Riby, J.E.; Fujisaws, T.; and Ferreia, S.H. (2001). Physico-chemical stability of natural or pre-sweetened frozen passion fruit juice. Braz. Arch. Biol. Technol., 42(4): 393–396.

20. Silva, F.V.M. and Gibbs, P. (2001). Alicyclobacillus acidoterrestris spores in fruit products and design of pasteurization processes. Trends Food Sci. Technol., 12: 68–74.

21. Rezk, A.S. (2003). Technological and microbiological studies on some juices. M.Sc. Thesis, Fac. Agric. Moshtohor, Benha Univ., Egypt.

22. Tran, M.T.T. and Farid, M. (2004). Ultraviolet treatment of orange juice. Innov. Food Sci. Emerg. Technol., 5: 495–502.

23. Guerrero-Beltrán, J.A. and Barbosa-Cánovas, G.V. (2005). Reduction of Saccharomyces cerevisiae, Escherichia coli and Listeria innocua in apple juice by ultraviolet light. J. Food Process Eng., 28: 437–452.

24. Tournas, V.H.; Heeres, J.; and Burgess, L. (2006). Moulds and yeasts in fruit salads and fruit juices. J. Food Microbiology., 23: 684–688.

25. Chang, S.S. and Kang, D.H. (2004). Alicyclobacillus spp. in the fruit juice industry: History, characteristics, and current isolation/detection procedures. Crit. Rev. Microbiol., 30: 55–74.

26. Komitopoulou, E.; Boziaris, I.S.; Davies, E.A.; Delves-Broughton, J.; and Adams, M.R. (1999). Alicyclobacilus acidoterrestris in fruit juices and its control by nisin. Int. J. Food Sci. Technol. 34: 81–85.
27. Karagözlü, N. (2004). Meyve sularında bozulma etmeni: Alicyclobacillus acidoterrestris. J. Agric Fac. HR. U. 8: 15–21.
28. Matsubara, H., Keiichi, G., Terumi, M., Kaoru, M., Masaharu, I., Motohiro, N. and Kazuhide, Y. (2002). Alicyclobacillus acidiphilus sp. nov., a novel thermo-acidophilic, omega-alicyclic fatty acid-containing bacterium isolated from acidic beverages. Int. J. Syst. Evol. Microbiol., 52: 1681–1685.
29. Martorell, P., Stratford, M., Hazel, S., Teresa, F and Amparo, Q. (2007). Physiological characterization of spoilage strains of Zygosaccharomyces bailii and Zygosaccharomyces rouxii isolated from high sugar environments. Int. J. Food Microbiol., 114: 234–242.
30. Arias, R., Jacqueline, K., Lorrie, M., Renee, M., and Mickey, E. (2002). Yeast species associated with orange juice: Evaluation of different identification methods. Appl. Environ. Microbiol., 68(4): 1955–1961.
31. Steels, H., James, S.A., Bond, C.J., Roberts, I.N., Stratford, M. (2002). Zygosaccharomyces kombuchaensis: The physiology of a new species related to the spoilage yeasts Zygosaccharomyces lentus and Zygosaccharomyces bailii. FEMS Yeast Res., 2: 113–121.
32. Oliveira, K.F.; Malavolta, L.; Souza, C.S.; Vicente, E.J.; and Laluce, C. (2006). Pectinolytic activity secreted by yeasts isolated from fermented citrus molasses. J. Appl. Microbiol. 100: 633–640.
33. Khalifa, S.G.; Khalifa, B.; Widad, B.E.A.; Saleha, E.E.N.; and Ali, Z. (2005). Microbiological quality of fruit juices sold in Tripoli–Libya. J. Food Control., 16: 855–858.
34. Song, H.; Byun, M.; Jo, C.; Lee, C.; Kim, K.; and Kim, D. (2007). Effects of gamma irradiation on the microbiological, nutritional, and sensory properties of fresh vegetable juice. Food Control., 18: 5–10.
35. Tsige, K.; Tsegaye, G.; and Ketema, B. (2008). Microbiological safety of fruit juices served in cafes/resturants, Jimma Town, Southwest Ethiopia. Microbiological Safety of Fruit Juices Served, Ethiop J. Health Sci., 18(3): 95–100.
36. Ameyapoh, Y.; de Souza, C.; and Traore, A. S. (2008). Hygienic quality of traditional processing and stability of tomato (Lycopersicon esculentum) puree in Togo. Bioresour. Technol., 99(13): 5798–5803.
37. Tasnim, F.; Anwar, H.; Nusrath, S.; Kamal, H.; Lopa, D; and Formuzul, H. (2010). Quality assessment of industrially processed fruit juices available in Dhaka city, Bangladesh. Maik Nutr., 16(3): 431–438.

38. Rahman, T.; Hasan, S.; and Noor, R. (2011). An assessment of microbiological quality of some commercially packed and fruit juice available in Dhaka City: A comparative study. S. J. Microbiol., 1(1): 13–18.

39. Oranusi, S.U.; Braide, W.; and Nezianya, O.H. (2012). Microbiological and chemical quality assessment of some commercially packed fruit juices sold in Nigeria. J. Biol. Sci., 2(1): 1–6.

40. Duarte, I. F., Delgadillo, I. and Gil, A. M. (2006). Study of natural mango juice spoilage and microbial contamination with Penicillium expansum by high resolution 1H NMR spectroscopy. Food Chem., 96(2): 313–324.

41. Alia M.A. El-Garably (2011). Quality characteristics and storage changes of mixed mango and papaya nectar. J. Home Econ., 2(1).

42. Renuka, B.; Kulkarni, S.G.; Vijayanand, P.; and Prapulla, S.G. (2009). Fructooligosaccharide fortification of selected fruit juice beverages: Effect on the quality characteristics. LWT—Food Science and Technology, 42: 1031–1033.

43. Roberfroid, M.B. (2007). Prebiotics: The concept revisited. J. Nutr., 137: 830S–837S.

44. Nualkaekul, N. and Charalampopoulos, D. (2013). Survival of probiotics in fruit juices. SWU Sci. J., 29(2): 238–262.

45. Tamime, A.Y. (2005). Probiotic Dairy Products, vol. 1. Oxford, UK: Blackwell Publishing.

46. Choi, E.A. and Chang, H.C. (2015). Cholesterol-lowering effects of a putative probiotic strain Lactobacillus plantarum EM isolated from kimchi. LWT - Food Science and Technology, 62: 210–217.

47. Tomoko, K.; Yamano, T.; Fukushima, Y.; Yoshino, H.; Mito, N.; and Sato, K. (2007). Effect of L. Johnsonii La1 on immuno function and serum albumin in aged and malnuourished aged mice. Nutrition, 23: 342–350.

48. Lollo, P.C.B.; Moura, C.S.; Morato, P.N.; Cruz, A.G.; Castro, W.F.; Betim, C.B.; Nisishima, L.; Faria, J.A.F.; Junior, M, M.; Fernandes, C. O.; and Amaya-Farfan, J. (2013). Probiotic yoghurt offers higher immune-protection than probiotic whey beverage. Food Res. Int., 54, 118–124.

49. El-Nezami, H.; Mykkanen, H.; Kankaanpaa, P.; Salminen, S.; and Ahokas, J. (2000). Ability of Lactobacillus and Propionibacterium strains to remove aflatoxin B1 from chicken duodenum. Journal Food Protection, 63(4): 549–552.

50. Haghshenas, B.; Abdullah, N.; Nami, Y.; Radiah, D.; Rosli, R.; and Khosroushahi, A.Y. (2014). Different effects of two newly-isolated

probiotic Lactobacillus plantarum 15HN and Lactococcus lactis subsp. Lactis 44Lac strains from traditional dairy products on cancer cell lines. Anaerobe, 30: 51–59.

51. Simakachorn, N.V.; Pichaipat, P.; Rithipompaisarn, C.; Kaew, K.; Tongpradit, P., and Vararithya, W. (2000). Clinical evaluation of the addition of lyophilized heat killed Lactobacillus acidophilus LB to oral dehydration therapy in the treatment of acute diarrhea in children. J. Pediatr. Gastroenterol. Nutr., 30(1): 68–72.
52. Sheela, T.; Kulothugan, S.; and Saravanamuthu, R. (2011). Antagonistic action of synbiotic carrot juice against diarrhoea causing organisms. Indian J. Appl. Pure Biol., 26(1): 15–22.
53. Labib, A.A.S. (1992). Chemical and technological studies on juices of some Egyptian fruits. Ph.D. Thesis, Fac. of Agric., Ismailia, Suz Canal Univ., Egypt.
54. Laratta, B.; Loiudice, R.; Giovane, A.; Quagliuolo, L.; Servillo, L.; and Castado, D. (1995). Thermostability of three pectinesterase isoenzymes in tomato fruit. Food Chem., 52(4): 415–418.
55. Ahmed, A.E.M. (2000). Effect of thermal processing on enzyme activity in some fruit juices. M.Sc. Thesis, Food Sci. Dep. Fac. of Agric.; Moshtohor, Zagazig Univ., Egypt.

Chapter 2

Grape and Winery By-product Extracts as Natural Preservatives in Sustainable Food Systems

Veronica Sanda Chedea,[a]* Liliana Lucia Tomoiagă,[a] Monica Trif,[b] Corina Ioana Bocsan,[c] and Raluca Maria Pop[c*]

[a]*Research Station for Viticulture and Enology (SCDVV Blaj), Romania*

[b]*Centre for Innovative Process Engineering (CENTIV) GmbH, Germany*

[c]*Department of Pharmacology, Toxicology and Clinical Pharmacology, "Iuliu Hatieganu" University of Medicine and Pharmacy, Victor Babes, No 8, 400012 Cluj-Napoca, Romania*

chedeaveronica@yahoo.com, raluca_parlog@yahoo.com

*These authors have contributed equally.

Industry and science are also very keen to extract natural alternatives to conventional preservatives from a wide variety of plants and fruits, or from the plant, fruits, and vegetables based residual biomass. Often attempts are also made to create a combination of preservatives and antioxidants with the help of a single extract to obtain a product that is both microbiologically harmless and visually

Natural Food Preservatives
Edited by Sourish Bhattacharya and Heba Hassan Abd-El Azim Salama

ISBN 978-981-4968-64-5 (Hardcover), 978-1-003-36776-5 (eBook)
www.jennystanford.com

or tastefully appealing. Grape-based natural and grape by-products extracts have a high potential as preservatives for food applications. Large amounts of grape by-products are normally treated as waste and generally disposed of in landfills but could potentially be used in the production of natural food preservatives. Grape pomace and grape seeds represent a viable alternative to synthetic preservatives that are commonly used to inhibit and or delay the processes that cause food spoilage and enhance the quality, safety, and shelf stability of foods.

2.1 Introduction

The aspects that shape the food environment are of biophysical and economic, cultural and political, and technological and physiological nature and they are very much linked together [1]. This interconnectedness can lead to feelings of helplessness when trying to change this environment [1]. This aggregate of factors evolved in time by interactions that reinforced each component as well as the relationship between them, and so changes in food systems will only occur if all of these elements are realigned with one another [1]. It is generally known and accepted that a shift from industrialized and input-intensive food production to sustainable production and from diets based mostly on ultra-processed and energy-dense foods to healthier consumption is needed [1]. In this context, the shift is only possible if everything, through the entire food chain, from farm to fork, changes at the same time, and if this is not possible, at least in an organized way at each level for the whole system [1].

The food chain generates besides the final product also by-products that contain a substantial proportion of nutrients originating from the input materials [2]. The high amounts of resulting by-products, their impact on the environment, and their nutritional composition make these matrices a significant topic for careful recovery and valorization [2].

Also, this represents a great opportunity, not only to reduce greenhouse gas emissions and fossil fuels [3] but also to contribute to the generation of multiple valuable ideas on how to valorize vegetable waste into different value-added by-products. Many research groups around the world are aiming to provide scientifically needed proofs and valuable ideas for optimizing the use of waste generation in various sectors such as animal feedstock, food industry, cosmetics, and pharmaceuticals.

According to the Food and Agriculture Organization of the United Nations (FAO), fruits and vegetable waste production is considered the highest compared with all types of foods, reaching up to 60%. Among fruits, grapes (*Vitis vinifera*) are the most cultivated fruit crops around the world. The annual production was estimated at 77.8 million tons in 2018, with 57% representing the wine grape production, 36% the table grape production, and 7% the dried grape production [4, 5]. The most important wine producers are Spain, China, France, Italy, and Turkey [4].

Within the wine processing industry, the main waste by-products are represented by grape pomace with 13–30% of grapes' total weight [6, 7]. Grape stems, grape leaves, and wine lees are other important by-products of waste (Fig. 2.1).

Currently, these by-products are considered waste disposal and in a small percentage, they are used as fertilizer or animal feed. Even though part of winery by-products are valorized worldwide (around 30–40% of total waste production), there is much room for the development of this sector [8]. The by-products from winery (grape pomace, grape seeds, and stems) alone or in combination are gaining interest as alternative additives to sulfur dioxide, with antioxidant and antimicrobial activity, and one of the preservatives most used in the wine industry actually [9].

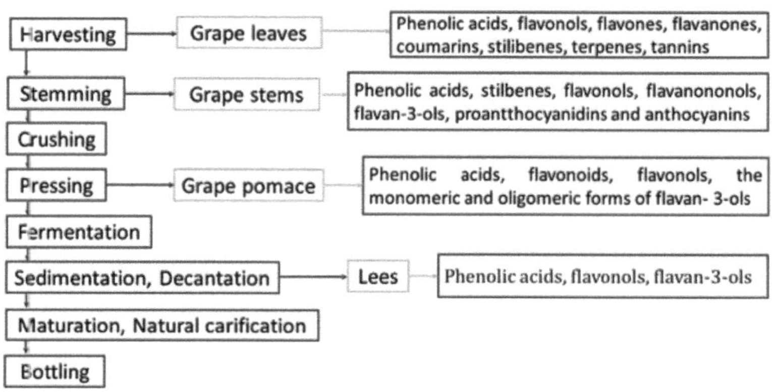

Figure 2.1 Main waste by-products resulted during the wine processing.

Therefore, the main aim of this chapter is to offer a complete overview regarding the perspective of valorizing these by-products, especially for their potential as food preservatives. For this reason, their chemical composition and their potential antimicrobial effects will be discussed to prove their usefulness in this field.

2.2 Chemical Composition of Main Grape By-products

The composition of the grape waste by-products is highly dependent on several factors such as grape variety, cultivars, conventional or organic cultivation, processing method, harvesting period, storage conditions, soil, region, climate conditions, and many other factors that automatically will induce a great variability among their chemical composition. Since the by-product collection for the retrieval of phenolics is of importance, and therefore, has to be taken into consideration the fact that diverse plant organs or tissues can be characterized by different phenolic composition [10, 11].

2.3 Chemical Composition of Grape Leaves

The general composition of grape leaves is highly dependent on the previously mentioned factors as well. The moisture content was found to vary between 7.39% and 8.20%, the oil content between 1.06% and 2.06%, the protein content between 9.38% and 20.3%, the carotenoid content between 13.92 and 60.46 ug/g, total phenolic content between 1298 and 1748 mg gallic acid equivalents (GAE)/100 g, and total flavonoid content between 5346 and 11239 mg quercetin equivalents/100 g [12]. Also, important quantities of minerals such as P, K, Ca, Mg, Na, and Fe were reported [12]. Other phytochemical compounds identified in the literature studies regarding the grape leaves composition include a large variety of organic acids (ascorbic, citric, linoleic, fumaric, oxalic, malic, and tartaric acids), terpenes, tannins, and vitamin E [13].

The major grape polyphenols identified in leaves are gallic acid, dihydroxybenzoic acid, catechin, dihydroxybenzene, caffeic acid and rutin, quercetin, and apigenin derivatives (Fig. 2.2) [12].

Hydroxybenzoic acids

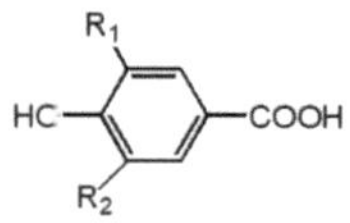

Compound	R1	R2
Gallic acid	OH	OH
p-Hydroxybenzoic acid	H	H
Protocatechuic acid	OH	H
Syringic acid	OCH_3	OCH_3

Hydroxycinnamic acids

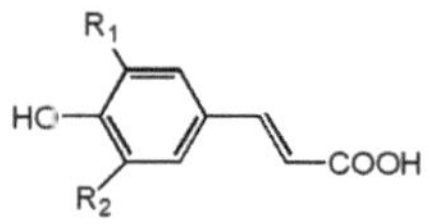

Compound	R1	R2
Caffeic acid	OH	H
p-coumaric acid	H	H
Ferulic acid	OCH_3	H
Sinapic acid	OCH_3	OCH_3

Flavanols

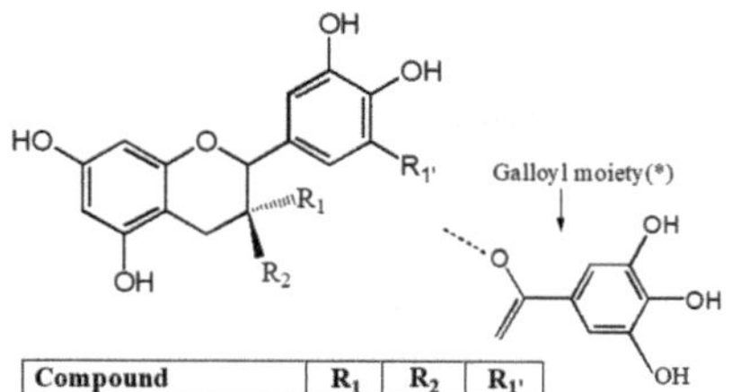

Compound	R_1	R_2	$R_{1'}$
(+)-Catechin	H	OH	H
(-)-Epicatechin	OH	H	H
(-)Epigallocatechin	OH	H	OH
(-)-Epicatechingallat	*	H	H
(-)-Epigallocatechingallat	*	H	OH

Stilbenes

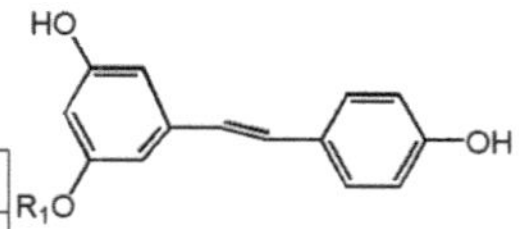

Compound	R_1	R_2
trans-Resveratrol	H	H
trans-Polydatin	glucose	OH

Flavonols

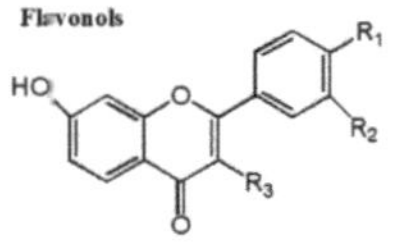

Compound	R_1	R_2	R_3
Kaempferol	OH	H	H
Quercetin	OH	OH	H
Isorhamnetin 3-O-glucoside	OH	OCH_3	glucose
Kaempferol 3-O-glucoside	OH	H	glucose
Quercetin 3-O-glucoside	OH	OH	glucose
Quercetin 3-O-galactoside	OH	OH	galactose
Quercetin 3-O-glucuronide	OH	OH	glucuronic acid
Quercetin 3-O-rhamnoside	OH	OH	rhamnose
Rutin	OH	OH	rutinoside

Anthocyanins

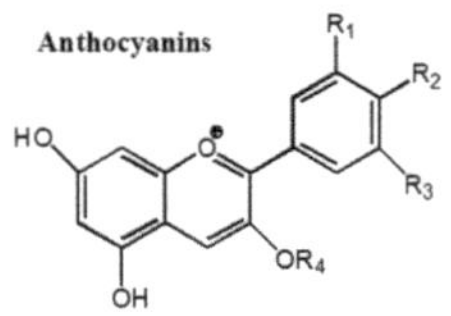

a= H
b=glucose
c=p-coumaroyl-glucose
d=acetyl-glucose

Compound	R_1	R_2	R_3	R_4
Malvidin	OCH_3	OH	OCH_3	a
Malvidin 3 -O-glucoside	OCH_3	OH	OCH_3	b
Malvidin 3 -O-acetylglucoside	OCH_3	OH	OCH_3	d
Malvidin 3-O-p-coumaroylglucoside	OCH_3	OH	OCH_3	c
Peonidin	OCH_3	OH	H	a
Peonidin 3 -O-glucoside	OCH_3	OH	H	b
Peonidin 3 -O-acetylglucoside	OCH_3	OH	H	d
Peonidin 3-O-p-coumaroylglucoside	OCH_3	OH	H	c
Petunidin	OCH_3	OH	OH	a
Petunidin 3 -O-glucoside	OCH_3	OH	OH	b
Petunidin 3 -O-acetylglucoside	OCH_3	OH	OH	d
Petunidin 3-O-p-coumaroylglucoside	OCH_3	OH	OH	c
Cyanidin	OH	OH	H	a
Cyanidin 3 -O-glucoside	OH	OH	H	b
Cyanidin 3-O-p-coumaroylglucoside	OH	OH	H	d
Delphinidin	OH	OH	OH	a
Delphinidin-glucoside	OH	OH	OH	b
Delphinidin 3-O-acetylglucoside	OH	OH	OH	d

Procyanidins

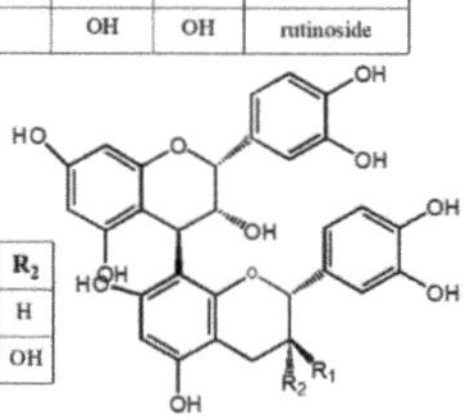

Compound	R_1	R_2
Procyanidin B1	OH	H
Procyanidin B2	H	OH

Figure 2.2 Chemical structures of the common phenolics identified in grape by-products [15].

2.4 Chemical Composition of Grape Stems

Grape stems, which represent the skeleton of grapes (composed of shoot, rachis, peduncle, and pedicel), also present a great variability in general composition in accordance with the previously mentioned factors. Thus, the water content of stems was found to vary between 55% and 80%, the cellulose and hemicellulose between 12% and 38% dry matter (DM), the lignin between 13% and 47% DM, the proteins between 5% and 11% DM, ashes around 6.9% DM, the acids between 13.5 and 15 g/kg fresh weight (FW), sugars between 1.8 and 3.7 g/100 g stem FW, and total polyphenols from 348 to 38,400 mg GAE/100 g DM [14]. Minerals such as K, Ca, Mg, Zn, and Na were also identified.

As observed, grape stems have significant amounts of polyphenols, generally with higher amounts in the red varieties [14]. Among polyphenols, phenolic acids, stilbenes, flavonols, flavan-3-ols, proanthocyanidins, and anthocyanins were reported [14, 16]. The main phenolic acids identified in grape steams were caftaric, coutaric, gallic, coumaric, caffeic, syringic, and ferulic acids [16]. The main identified stilbenes were trans-resveratrol and ε-viniferin [16]. Among flavonols, there were derivatives of quercetin, kaempferol, myricetin, and isorhamnetin, mainly glucosides, rhamnosides, and rutinosides [16]. Astilbin and engeletin were the main flavanonols. The flavan-3-ols and proanthocyanidins were mainly represented by catechin and B1 and B3 dimers while the anthocyanins by malvidin derivatives (Fig. 2.2) [16].

2.5 Chemical Composition of Grape Pomace

Grape pomace mainly containing seeds, skins, and grape stems represents a rich source of bioactive compounds. Its general composition was also reported to have great variability, depending on the previously mentioned factors, especially on grape varieties and fermentation steps in the processing method. Thus, the ash was reported between 1.79 and 9.10 g/100 g, the proteins between 3.57 and 14.17 g/100 g, the fats between 1.14 and 13.90 g/100 g, total dietary fibers between 17.28 and 88.7 g/100 g, the carbohydrates between 12.20 and 40.53 g/100 g, and the total polyphenols between 0.28 and 8.70 g/100 g [17]. Minerals such as Na, K, Mg, Ca, Mn, Fe, Zn, Cu, and P were also identified.

Regarding polyphenolic compounds, characteristics for grape pomace are the phenolic acids, flavonoids, flavonols, the monomeric and oligomeric forms of flavan-3-ols found in seeds, and anthocyanins found exclusively in the red varieties of grape skins [7] (Fig. 2.2). Phenolic acids are represented by hydroxybenzoic acids such as quinic, protocatechuic, and vanillic acids and by hydroxycinnamic acids such as feruloyltartaric, *p*-coumaric, caffeic, caftaric, chlorogenic, ferulic, sinapic, coutaric, and fertaric acids [15, 18]. Flavonols are represented mainly by myricetin, quercetin, kaempferol, and isorhamnetin derivatives. The flavan-3-ols and the tannins by epicatechin, catechin, and B1, B2, B3, and B4 procyanidins, while the anthocyanins by cyanidin, peonidin, and malvidin derivatives [15, 18]. The stilbenes by trans-resveratrol, e-viniferin and trans-piceid and piceatannol [15, 18] (Fig. 2.2).

2.6 Chemical Composition of Lees

The composition of lees is mainly represented by the microorganisms, specifically the yeasts that are produced during alcoholic fermentation. It also contains, in smaller amounts, bacteria, phenolic compounds, plant cell remnants, tartaric salts, and inorganic matter [19]. As expected, the composition of wine lees also depends on several factors. Besides the above-mentioned factors, the time necessary for the aging process in the wood barrels is also important. This factor is important since the wine aging process on lees is considered a traditional oenological technique that implies placing the wines on the lees and the grape solids [20]. Thus, a part of phenolic compounds (26 to 546 mg GAE/g DM) remains during the winemaking process in several by-products such as lees [20]. Also, important quantities of total flavonoids (around 984 mg GAE/100 g DM) and total anthocyanins (383 mg of cyanidin-3-glucoside equivalents/100 g of DM) were reported [20].

The profile of phenolics characterises for the wine comprising less concentration of flavonols such as quercetin, quercitrin, myricetin, and kaempferol derivatives, flavan-3-ols such as catechin, epicatechin, and procyanidin B2, anthocyanidins such as petunidin, peonidin, delphinidin, cyanidin, and malvidin derivatives, and phenolic acids such as caffeic, p-coumaric, gallic, and vanillic acids [20] (Fig. 2.2).

The search has shown differences in phenolic content between grape pomace and grape seeds, mainly stemming from tissue differences. Due to fluctuating environmental conditions and developmental stages, the fruits synthesize and localize the phenolic compounds differently, in various internal structures (i.e., cells, organelles). Grape pomace has the highest content of total phenols, tannins, and carotenoids, and grape seeds have the highest content of proanthocyanidins, and they have been proven to be ideal candidates as natural antioxidants and antimicrobials for food applications [21].

2.7 Need for Naturally Occurring Preservatives for Food

Many industrially processed foods contain questionable synthetic additives such as certain preservatives. The shelf life can be extended by using some herbs, spices, essential oils, or by-products. Incidentally, the imprint "without preservatives" only means that no preservatives have been added in the legal sense. However, this does not mean that it does not contain any substances with a preservative effect. Some antioxidants are also used for preservation. They protect food from spoilage due to atmospheric oxygen (oxidation), light, or other substances. Oxidation can be seen when the food turns brown. Acidifiers also extend the shelf life of food because many undesirable microorganisms are sensitive to acids. Acidity regulators also support the effect of preservatives, as these often only work optimally at certain pH values. Preservatives and substances with a preservative effect (antioxidants and acidulants) are intended to extend the shelf life of food by preventing the growth of microorganisms such as mold, putrefactive agents, and fermentation agents. In fact, toxins produced by yeasts can cause serious health risks. Molds are carcinogenic and toxic to nerves, kidneys, and liver, can cause deformities, and weaken the immune system. Chemically synthesized preservatives such as benzoate, citric acid, nitrate, nitrite, propionate, sorbate, sulfites, or tartaric acid are commonly used for the microbial growth control being authorized by regulatory agencies [22]. However, long-term use of synthetic preserving agents has been shown to cause serious health issues, such as liver and kidney injury, diseases of gastrointestinal function, asthma, some cancers, and multiple allergies [22, 23]. Because contemporary consumers are more and more aware of this fact, the food producers as well as the consumers demand the use of natural preservatives instead of synthetic ones for a healthier diet with a less processed food [22, 24, 25].

Another issue that has to be taken into consideration when we discuss the need for naturally occurring antimicrobial agents is the fact that over the past decades, globally, the antibiotic resistance is a serious health challenge [22]. It was found that more than 70% of the pathogenic bacteria are resistant to antibiotics [26] and this high percentage of resistant microorganisms is an important concern for public health.

2.8 Antimicrobial Effect of Polyphenols

As presented earlier, the grape and wine products contain important amounts of different classes of polyphenols, compounds with a well-documented antimicrobial activity (Fig. 2.3) [27] besides the other health-promoting effects – antioxidant, anti-inflammatory, antihypertensive, anticancer, antidiabetic, anti-obesity, and cognitive enhancing [28–30]. This is the background for the multitude of studies demonstrating the role of polyphenols as promising potential natural antimicrobial food preservatives for inactivating spoilage and pathogenic microorganisms and improving microbial protection [22].

Natural preservatives with antioxidant activities are preferred by the consumer. The extracts are therefore particularly suitable for protecting food from harmful bacteria or can be incorporated in biodegradable active films and packaging for food applications. However, the right dishes that are not influenced by the pomace essence either visually or in terms of taste still need to be found.

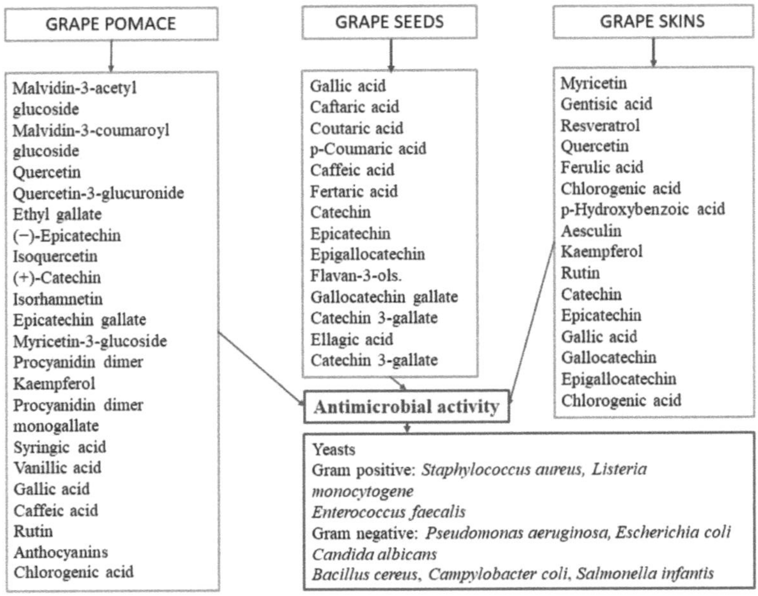

Figure 2.3 Grape by-products antimicrobial activity.

The microbial growth of the foodborne pathogens such as *Staphylococcus aureus*, *Listeria innocua*, and *Escherichia coli*, in the absence and the presence of a red wine pomace seasoning (RWPS), was assessed for 34 h at 37°C [31]. The study has shown that this grape pomace product had bactericidal effects against *S. aureus* and *L. innocua* at a concentration of 40 g/L and bacteriostatic effects against *E. coli* at the same concentration and against *S. aureus* at half concentration (20 g/L) [31]. In addition, RWPS delayed the growth of *E. coli* and *L. innocua* at 40 g/L, extending the lag phase and lowering the maximum growth rate of these two pathogens [31]. In another study, the potential of a polyphenolic grape seed extract (PE) against *E. coli* and *Brevibacterium linens* for use as a natural antibacterial agent was evaluated [32]. After 48 h incubation on a solid medium, PE had a bactericidal effect on the Gram-positive *B. linens* and a reduction of the microbial growth of *E. coli* [32]. Like in other similar studies, it was observed that the antimicrobial effect of the polyphenols is strain and dose dependent and that the extracts have shown stronger antimicrobial potency than individual polyphenols [22, 24, 32]. The electronic and charge properties of polyphenols as well as their lipophilicity [33] influence greatly the interaction between these compounds and the bacterial cell surface [24], and thus they exert their antimicrobial action [22]. The interaction of polyphenols with the bacterial cell surface is followed by the penetration of the pathogens' membrane and their reaction with the cytoplasm or cellular proteins [34].

2.9 Grape and Wine By-products as Food Preservatives

Different grape and wine by-products were studied for their use as food preservatives (Table 2.1).

2.9.1 In Meat

Known for their high content in procyanidin dimers and trimers as well as in epicatechin, catechin, and epicatechin gallate [43], the grape seed extracts were reported to control the growth of pathogenic bacteria in beef meat products [35]. The controlled growth of the pathogenic bacteria *L. monocytogenes* and *A. hydrophila* and the significantly reduced counts of *E. coli* O157:H7 and *S. typhimurium* were registered by Ahn *et al.* (2007) in cooked beef samples treated with 1% grape seed extracts [35] (Table 2.1).

Maybe, the most important by-product of the wine industry, grape pomace was also studied as a meat preservative. Thus, different grape pomace extracts obtained from five Turkish grapevine varieties were tested for their antimicrobial activity in beef patties at 0% (control), 1%, 2%, 5%, and 10% concentrations for different storage periods (0, 12, 24, and 48 h) [36] (Table 2.1). Generally, the microorganisms count was lower by all the extracts concentrations during all the storage periods, the best result being registered for the 10% concentration for all the times tested [36]. Moreover, 5% grape pomace extracts of the Emir, Gamay, and Kalecik Karasi varieties in beef patties also inhibited the foodborne pathogens including Enterobacteriaceae and coliform bacteria as well as spoilage microorganisms including yeasts and lipolytic bacteria [36]. The Enterobacteriaceae growth was also completely inhibited in a beef meat homogenate prepared by homogenizing commercially

Table 2.1 Different grape and wine by-products studied for their usage as food preservatives

Grape by-product/reference	Concentration/ used food	Investigated microorganism	Observation
Grape seed extract (GSE) / [35]	1%/cooked beef	Reduced *E. coli* O157:H7 and *Salmonella typhimurium* and controlled the growth of *L. monocytogenes* and *Aeromonas hydrophila*	–
Grape pomace extracts (GPE)/[36]	0%, 1%, 2%, 5%, and 10%/100 g ground beef	Enterobacteriaceae, Coliform bacteria, *S. aureus, Salmonella*, enterococci, total aerobic mesophilic bacteria, total aerobic psychrotrophic bacteria, yeasts and molds, Micrococcaceae	– GPE had the antimicrobial effect – Prevented the patties deterioration – At 5% and 10% inhibited some foodborne pathogens
Whole wine pomace product (WWPP), seed-free WP, termed skin wine pomace product (SkWPP), the third from the isolated seeds, termed seed wine pomace product (SdWPP)/[37]	2%/beef meat homogenates in sterile water (1:3)	Total aerobic mesophilic bacteria (TAMB), total yeasts and molds, lactic acid bacteria (LAB), Enterobacteriaceae	– Complete inactivation of LAB after 32 h of incubation – No Enterobacteriaceae growth – SkWPP had a bactericidal effect – WWPP had bacteriostatic activity against TAMB and LAB growth (after 15 and 22 h) – WWPP delayed Enterobacteriaceae growth and decreased the final counts

Grape by-product/reference	Concentration/ used food	Investigated microorganism	Observation
Red wine grape pomace (RWSP)/[31]	2%/beef patties	Total aerobic mesophilic bacteria (TAMB) lactic acid bacteria (LAB)	– RWPS delayed TAMB and LAB growth in low-salt beef patties – RWPS increased the stability of normal patties (2% salt), reduced the use of allergenic preservatives (e.g., sulfites)
Red grape pomace extract (RGPE)/[38]	1000 mg red grape pomace extract/1 kg meat in lamb patties	Total aerobic mesophilic bacteria, psychrotrophic counts, Enterobacteriaceae, lactic acid bacteria, *Salmonella* spp., and *L. monocytogenes* microbial	– RGPE had an antimicrobial effect – Significantly lowered the mesophiles and psychrotrophics after 7-day storage – PGPE had no effect against Enterobacteriaceae and LAB
Red grape pomace extract (RGPE)/[39]	2% and 4%/carp (*Cyprinus carpio*) fillet	Total viable counts (TVC) and psychrophilic total count (PTC)	– Best results with 2% RGPE over 3 days and 4% RGPE over 6 days – 4% RGPE had the best preservation condition
Red Merlot and White Zeleny grape pomace/[40]	6%, 10%, and 15%/ bread	Total molds	– The effect increased with the concentration. The red grape pomace had a better action than the white one

(Continued)

Table 2.1 (*Continued*)

Grape by-product/reference	Concentration/ used food	Investigated microorganism	Observation
White grape pomace/[41]	5%/rice gluten-free pasta	The total plate count, yeast, and mold count	– The microorganisms were significantly lowered on the first and fifth days of storage when compared with the control samples
Lyophilized grape seed extract (LGSE) **Lyophilized grape seed extract and colloidal silver complex (LGSE-CSC)** **Lyophilized stem extract (LSE)** **Lyophilized stem extract and colloidal silver complex (LSE-CSC)/[9]**	0.5 g/L LGSE, 0.5g/L LGSE-SC 0.5 g/L LSE 0.5 g/L LSE-CSC/ Airén grapes (*V. vinifera*) from 2016 vintage must	Total yeasts, lactic acid, acetic bacteria counts	– Extracts had a significant effect against investigated microorganisms similar to that of SO_2 (control). The wines with stem extract and colloidal silver had the lowest final counts
Grape seed extract (GSE)/[42]	0%, 0.06%, 0.12%, 0.23%, 0.45%, 0.9%, 1.8%, 3.6%/apple juice	*Alicyclobacillus acidoterrestris* DSM 3922 vegetative cells and spores	Apple juice with 3.6% GSE had the best microorganism counts inhibition

available beef in sterile water (1:3) by a seed-free red pomace [37] (Table 2.1). This grape pomace product had as well bactericidal effects against total aerobic mesophilic bacteria (TAMB) and lactic acid bacteria (LAB) [37]. The same study shows that the whole red grape pomace presented bacteriostatic activity against the three microorganism groups tested, whereas the product obtained only from grape seed promoted TAMB and LAB growth but delayed the Enterobacteriaceae proliferation [37].

The meat industry is pursuing new solutions for reducing the meat products' sodium content without reducing their shelf life [31]. Seasonings like salt alternatives are far more valued than chemical preservatives and often allow the addition of interesting nutrients [31]. In this context, the ability of a seedless red wine pomace seasoning (RWPS), to prevent the beef patties' spoilage development for various salt concentrations (2%, 1.5%, and 1%) and kept at 4°C was studied [31] (Table 2.1). Total aerobic mesophilic and lactic acid bacteria growth, especially in low salt samples, was slowed by the incorporation of 2% w/w RWPS [31]. The growth delaying of TAMB and LAB in patties with low salt concentrations was done also when these were stored under modified atmosphere packaging and air packaging [31]. The patties containing a normal amount of salt (2%), had a higher stability to the microbial spoilage when they contained RWPS, and so the use of sulfites which are allergenic preservatives was reduced [31].

To check if grape pomace extract could preserve microbial quality, 1000 mg red grape pomace extract was mixed with 1 kg meat in lamb patties and total aerobic mesophilic bacteria, psychrotrophic counts, Enterobacteriaceae, lactic acid bacteria, *Salmonella* spp., and *L. monocytogenes* were assessed [9] (Table 2.1). The results of this test showed that microbial counts in lamb patties with aqueous extract were significantly lower than the controls for mesophiles and psychrotrophic after 7-day storage though the antimicrobial effect of extracts was not so evident for Enterobacteriaceae and LAB [9].

The presented results indicate that the grape pomace extracts as well as the dried grape pomace seasonings may be a good option for the extension of the food products' microbial shelf life and the inhibition of meat pathogens [31, 36, 37]. Moreover, the grape pomace seasoning had an interesting preservation capacity for reducing meat homogenates' salt and replacing sodium ascorbate, without losing their microbial stability, increased the concentrations of potassium and calcium, and helped the dietary fiber and phenolic compounds to be incorporated also [31].

2.9.2 In Fish

The effect of adding 0%, 2%, and 4% red grape pomace extract to common carp (*C. carpio*) fillets during their storage at 4°C was studied by determining total viable counts (TVC) and psychrophilic total count (PTC) [39] (Table 2.1). According to the analysis results, in treatments with 2% extracts after 3 days and treatments with 4% extract after 6 days, the shelf life was better than in control [39]. Overall, 4% extract administration gave the best preservation condition suggesting the use of this treatment in the preservation of fish fillet [39].

2.9.3 In Bread and Pasta

The effect of grape pomace in preserving and extending the shelf life of bread and pasta was also investigated [44]. The incorporation of different concentrations (6%, 10%, and 15%) of red Merlot and white Zeleny grape pomace in bread inhibited the growth of mold colonies, the effect increasing with the added pomace concentration [40]. The red grape pomace had a better action than the white one [40]. Not the same inhibitory effect was observed when *Aspergillus flavus*, *A. ochraceus*, *Penicillium nordicum*, and *P. verrucosum* were externally inoculated on the bread containing the grape pomace [40]. However, *A. flavus* was statistically significantly inhibited on the bread by the red grape pomace in the concentration of 6% after 3 days of incubation [40]. In rice gluten-free pasta where a white grape pomace was added in a concentration of 5%, the total plate count and yeast and mold count were significantly lower on the first and fifth days of storage when compared with the control samples [41].

2.9.4 In Beverages

Like in meat, the reduction of sulfites in winemaking is one of the current priorities, and for this reason the enological industry is trying to find natural additives with antioxidant and antimicrobial properties to replace or at least lower the use of these allergenic preservatives [9]. For this reason, Marchante et al. (2019) studied the use of aqueous grape seeds and stem extracts in combination with colloidal silver or alone for white wine production (Table 2.1). Their results demonstrated the antimicrobial activity of these extracts that was similar to that of sulfur dioxide [9]. The activity was higher in the wine in which the extracts were added in combination with the colloidal silver against the lactic and acetic bacteria, and the lowest final bacterial counts were found for the stem extract combined with colloidal silver [9].

Besides its antimicrobial activity, any preservative added to a food product has to pass the sensorial, color and taste influence test. As for a wine, its color and taste are of paramount importance, and the extracts must also pass these tests. So, in the case of Marchante et al.'s work, the wine containing the stems extract in combination with colloidal silver was most similar to those made with sulfur dioxide; the wines with extracts had greater scores on the fruity and floral attributes and the wine in which the grape seed extract had a slight bitterness received the worst evaluation [9].

A grape seed extract was added to apple juice in concentrations of 0%, 0.06%, 0.12%, 0.23%, 0.45%, 0.9%, 1.8%, and 3.6% to control vegetative cells and spores of *A. acidoterrestris* [42]. At the highest concentration, 3.6%, the microorganisms were inhibited showing that the grape seed extract can be used to improve the microbial stability of the apple juice [42].

2.10 Safety of Grape By-products as Food Preservatives

From the dietary point of view, the grape and winery by-products, containing important classes of polyphenols, dietary fibers, and polyunsaturated fatty acids, can be classified as nutraceuticals [45]. Compounds from natural sources are generally thought of as being healthy even at elevated doses, and therefore little evidence of the toxicity of natural plant by-products is found in the literature [22, 45].

The effects of dietary grape pomace were studied in animals [46–52], but in humans the effects and safety of grape and winery by-products as either food additives or dietary supplements are not clearly known and must be carefully studied. In this framework, besides the establishment of their therapeutic index when used to treat or as a complementary treatment in different diseases, it is also very important to evaluate their safety levels when used as food preservatives. Their function as preservatives in food is mainly influenced by their antimicrobial activity and antioxidant action. By consequence, an eventual cytotoxicity of these matrices is given by the modulation of their antioxidant/prooxidant balance. The oxidative potential of a polyphenolic grape seed extract, with the idea of using this extract as a nutritive supplement, was evaluated and the results showed that to some extent the grape seed extract tested, considered as an antioxidant nutritive supplement, may have prooxidant activity as well, depending on the dose, duration of administration, and other dietary components [32]. Grape seed proanthocyanidin extract (GPSE) at high doses (100 or 500 μg/mL) has been shown to have a prooxidant effect inducing cytotoxicity that is associated with increased apoptotic cell death [53].

Concerning the safety of the grape and winery by-products, the most studied matrix is the grape seed extract. For instance, Yamakoshi et al. (2002) demonstrated that in rats high doses of 2 and 4 g/kg of a proanthocyanidins rich extract does not induce acute and subchronic toxicity, that the lethal dose is greater than 4 g/kg after 14 days of administration, and that a dose estimated to be 240 times higher than the amount ingested by humans (2 g/kg) in the no-observed-adverse-effect level (NOAEL) during 90 days of a subchronic toxicity study [54]. Another subchronic toxicity study of GSE incorporated into chow and fed to rats for at least 3 months has indicated that a dietary concentration of approximately 1.78 g/kg body weight/day GSE in male rats and 2.15 g/kg body weight/day in female rats was a no-observed-adverse-effect level (NOAEL) [55]. Under the experimental conditions of a 2-month-repeated dosing oral toxicity trial evaluating the safety of increasing the dosage of grape seed powder (to 0.4, 4, 8, and 16 g/kg BW), administered subchronically to Wistar rats, the treatment appeared as highly safe even when administered at very high dosage and it could find potential applications in various biotic or abiotic stress-mediated multi-organ dysfunction [56]. The safety of the grape seed products has been proven also in humans (150 mg GSE per day, and 300 mg GSE per day for 4 weeks) suggesting that grape seed extract could be used as a nutraceutical for patients with the metabolic syndrome [57]. Officially, the safety and efficacy of the dry grape extract, a mixture of two extracts from seeds and skin of *V. vinifera* subsp. *vinifera*, was confirmed by the European Food Safety Authority (EFSA) Panel on Additives and Products or Substances used in Animal Feed (FEEDAP) at the maximum proposed use level (100 mg/kg complete feed) for all animal species except dogs, for which no safe dietary concentration could be established [58].

Finding the appropriate doses of different grape and winery by-products for healthy dietary use by intensive testing efforts to prevent any adverse side effects is necessary [22, 45].

2.11 Conclusion

The most studied grape and winery products in view of their use as food preservatives are the grape seeds and the grape pomace. Even though it was demonstrated that the whole grape pomace seasonings may be a good approach for the innovative food processing, this matrix is little explored compared with the extracts obtained from grape pomace, many more reports describing the use of these extracts as food additives. On the other hand, the extracts are demonstrated to have a better antimicrobial action compared with the single isolated compounds. The antimicrobial action of the grape and winery products is dependent on the dose and the pathogenic strain. In this context, the future challenge is to find the right dietary dose of each preservative by-product to prevent any side effects when used for human diet.

References

1. Mark, L., Sharon, F. (Eds.). Healthy and Sustainable Food Systems; Routledge, 2019.

2. Banaszewska, A.; Cruijssen, F.; Claassen, G.D.H.; van der Vorst, J.G.A.J. Effect and key factors of byproducts valorization: The case of dairy

industry. J. Dairy Sci. 2014, 97, 1893–1908, doi:10.3168/jds.2013-7283.

3. Rawdkuen, S.; Kaewprachu, P. Valorization of food processing by-products as smart food packaging materials and its application. doi:10.5772/intechopen.86245.
4. OIV; International Organisation of Vine and Wine. 2019 Statistical Report on World Vitiviniculture. doi:64/19/6835 [pii]\n10.1158/0008-5472.CAN-04-1678.
5. Ky, I.; Lorrain, B.; Kolbas, N.; Crozier, A.; Teissedre, P.L. Wine by-products: Phenolic characterization and antioxidant activity evaluation of grapes and grape pomaces from six different French grape varieties. Molecules 2014, 19, 482–506, doi:10.3390/molecules19010482.
6. Luchian, C.E.; Cotea, V.V.; Vlase, L.; Toiu, A.M.; Colibaba, L.C.; Răschip, I.E.; Nadăş, G.; Gheldiu, A.M.; Tuchiluş, C.; Rotaru, L. Antioxidant and antimicrobial effects of grape pomace extracts. BIO Web Conf. 2019, 15, 04006, doi:10.1051/bioconf/20191504006.
7. Marchante, L.; Gómez Alonso, S.; Alañón, M.E.; Pérez-Coello, M.S.; Díaz-Maroto, M.C. Natural extracts from fresh and oven-dried winemaking by-products as valuable source of antioxidant compounds. Food Sci. Nutr. 2018, 6, 1564–1574, doi:10.1002/fsn3.697.
8. Duque-Acevedo, M.; Belmonte-Ureña, L.J.; Cortés-García, F.J.; Camacho-Ferre, F. Agricultural waste: Review of the evolution, approaches and perspectives on alternative uses. Glob. Ecol. Conserv. 2020, 22, e00902.
9. Marchante, L.; Loarce, L.; Izquierdo-Cañas, P.M.; Alañón, M.E.; García-Romero, E.; Pérez-Coello, M.S.; Díaz-Maroto, M.C. Natural extracts from grape seed and stem by-products in combination with colloidal silver as alternative preservatives to SO_2 for white wines: Effects on chemical composition and sensorial properties. Food Res. Int. 2019, 125, 108594, doi:10.1016/j.foodres.2019.108594.
10. Kalli, E.; Lappa, I.; Bouchagier, P.; Tarantilis, P.A.; Skotti, E. Novel application and industrial exploitation of winery by-products. Bioresour. Bioprocess. 2018, 5, 46.
11. Veberic, R. The impact of production technology on plant phenolics. Horticulturae 2016, 2, 8.
12. Banjanin, T.; Uslu, N.; Vasic, Z.R.; Özcan, M.M. Effect of grape varieties on bioactive properties, phenolic composition, and mineral contents of different grape-vine leaves. J. Food Process. Preserv. 2021, 45, e15159, doi:10.1111/jfpp.15159.

13. Lacerda, D.S.; Costa, P.C.; Funchal, C.; Dani, C.; Gomez, R. Benefits of vine leaf on different biological systems. In Grape and Wine Biotechnology; InTech, 2016.

14. Blackford, M.; Comby, M.; Zeng, L.; Dienes-Nagy, Á.; Bourdin, G.; Lorenzini, F.; Bach, B. A review on stems composition and their impact on wine quality. Molecules 2021, 26, 1–40, doi:10.3390/molecules26051240.

15. Fontana, A.R.; Antoniolli, A.; Bottini, R. Grape pomace as a sustainable source of bioactive compounds: Extraction, characterization, and biotechnological applications of phenolics. J. Agric. Food Chem. 2013, 61, 8987–9003, doi:10.1021/jf402586f.

16. Souquet, J.M.; Labarbe, B.; Le Guernevé, C.; Cheynier, V.; Moutounet, M. Phenolic composition of grape stems. J. Agric. Food Chem. 2000, 48, 1076–1080, doi:10.1021/jf991171u.

17. Antonić, B.; Jančíková, S.; Dordević, D.; Tremlová, B. Grape pomace valorization: A systematic review and meta-analysis. Foods 2020, 9, 1627, doi:10.3390/foods9111627.

18. Chowdhary, P.; Gupta, A.; Gnansounou, E.; Pandey, A.; Chaturvedi, P. Current trends and possibilities for exploitation of grape pomace as a potential source for value addition. Environ. Pollut. 2021, 278, 116796, doi:10.1016/j.envpol.2021.116796.

19. Delgado De La Torre, M.P.; Priego-Capote, F.; Luque De Castro, M.D. Characterization and comparison of wine lees by liquid chromatography-mass spectrometry in high-resolution mode. J. Agric. Food Chem. 2015, 63, 1116–1125, doi:10.1021/jf505331f.

20. Jara-Palacios, M.J. Wine lees as a source of antioxidant compounds. Antioxidants 2019, 8, 45.

21. Pfukwa, T.M.; Fawole, O.A.; Manley, M.; Gouws, P.A.; Opara, U.L.; Mapiye, C. Food preservative capabilities of grape (*Vitis vinifera*) and Clementine Mandarin (*Citrus reticulata*) by-products extracts in South Africa. Sustainability 2019, 11, 1–15.

22. Ofosu, F.K.; Daliri, E.B.M.; Elahi, F.; Chelliah, R.; Lee, B.H.; Oh, D.H. New insights on the use of polyphenols as natural preservatives and their emerging safety concerns. Front. Sustain. Food Syst. 2020, 4, 525810.

23. Etemadi, A.; Sinha, R.; Ward, M.H.; Graubard, B.I.; Inoue-Choi, M.; Dawsey, S.M.; Abnet, C.C. Mortality from different causes associated with meat, heme iron, nitrates, and nitrites in the NIH-AARP Diet and Health Study: Population based cohort study. BMJ 2017, 357, doi:10.1136/bmj.j1957.

24. Bouarab Chibane, L.; Degraeve, P.; Ferhout, H.; Bouajila, J.; Oulahal, N. Plant antimicrobial polyphenols as potential natural food preservatives. J. Sci. Food Agric. 2019, 99, 1457–1474.
25. Rico, D.; Martín-Diana, A.B.; Barat, J.M.; Barry-Ryan, C. Extending and measuring the quality of fresh-cut fruit and vegetables: A review. Trends Food Sci. Technol. 2007, 18, 373–386, doi:10.1016/j.tifs.2007.03.011.
26. Harvey, A.L.; Edrada-Ebel, R.; Quinn, R.J. The re-emergence of natural products for drug discovery in the genomics era. Nat. Rev. Drug Discov. 2015, 14, 111–129.
27. Lal, A.F.; Singh, S.; F.C., F.; Bhatia, S. Potential of polyphenols in curbing quorum sensing and biofilm formation in Gram-negative pathogens. Asian Pac. J. Trop. Biomed. 2021, 11, 231.
28. Milner, J.; Goldberg, I. Functional Foods; Chapman and Hall: London, 1994.
29. Sun, H.; Wang, D.; Song, X.; Zhang, Y.; Ding, W.; Peng, X.; Zhang, X.; Li, Y.; Ma, Y.; Wang, R.; et al. Natural prenylchalconaringenins and prenylnaringenins as antidiabetic agents: α-Glucosidase and α-amylase inhibition and in vivo antihyperglycemic and antihyperlipidemic effects. J. Agric. Food Chem. 2017, 65, 1574–1581, doi:10.1021/acs.jafc.6b05445.
30. Araya-Cloutier, C.; Vincken, J.P.; van Ederen, R.; den Besten, H.M.W.; Gruppen, H. Rapid membrane permeabilization of Listeria monocytogenes and Escherichia coli induced by antibacterial prenylated phenolic compounds from legumes. Food Chem. 2018, 240, 147–155, doi:10.1016/j.foodchem.2017.07.074.
31. García-Lomillo, J.; Viegas, O.; Gonzalez-SanJose, M.L.; Ferreira, I.M.P.L.V.O. Influence of red wine pomace seasoning and high-oxygen atmosphere storage on carcinogens formation in barbecued beef patties. Meat Sci. 2017, 125, 10–15, doi:10.1016/j.meatsci.2016.11.009.
32. Chedea, V.S.; Braicu, C.; Socaciu, C. Antioxidant/prooxidant activity of a polyphenolic grape seed extract. Food Chem. 2010, 121, 132–139, doi:10.1016/j.foodchem.2009.12.020.
33. Rastija, V.; Nikolić, S.; Masand, V.H. Quantitative relationships between structure and lipophilicity of naturally occurring polyphenols. Acta Chim. Slov. 2013, 60, 781–789.
34. Kumar, Y.; Yadav, D.N.; Ahmad, T.; Narsaiah, K. Recent trends in the use of natural antioxidants for meat and meat products. Compr. Rev. Food Sci. Food Saf. 2015, 14, 796–812, doi:10.1111/1541-4337.12156.

35. Ahn, J.; Grün, I.U.; Mustapha, A. Effects of plant extracts on microbial growth, color change, and lipid oxidation in cooked beef. Food Microbiol. 2007, 24, 7–14, doi:10.1016/j.fm.2006.04.006.

36. Sagdic, O.; Ozturk, I.; Yilmaz, M.T.; Yetim, H. Effect of grape pomace extracts obtained from different grape varieties on microbial quality of beef patty. J. Food Sci. 2011, 76, M515–M521, doi:10.1111/j.1750-3841.2011.02323.x.

37. García-Lomillo, J.; González-SanJosé, M.L.; Del Pino-García, R.; Rivero-Pérez, M.D.; Muñiz-Rodríguez, P. Antioxidant and antimicrobial properties of wine byproducts and their potential uses in the food industry. J. Agric. Food Chem. 2014, 62, 12595–12602, doi:10.1021/jf5042678.

38. Andrés, A.I.; Petrón, M.J.; Adámez, J.D.; López, M.; Timón, M.L. Food by-products as potential antioxidant and antimicrobial additives in chill stored raw lamb patties. Meat Sci. 2017, 129, 62–70, doi:10.1016/j.meatsci.2017.02.013.

39. Hasani, S.; Hasani, M. Antimicrobial properties of grape extract on Common carp (*Cyprinus carpio*) fillet during storage in 4 °C. Int. J. Fish. Aquat. Stud. 2014, 1, 130–136.

40. Šporin, M. Improving the nutritional value and a shelf life of bread with the addition of grape pomace and olive cake (PhD thesis). Ljubljana, Slovenia: University Of Ljubljana, 2019.

41. Savla, H.; Yardi, V. Development of gluten-free pasta (Sevaiya) using grape pomace and assessing its quality and acceptability. doi:10.7763/IPCBEE.2016.V95.8.

42. Molva, C.; Baysal, A.H. Antimicrobial activity of grape seed extract on Alicyclobacillus acidoterrestris DSM 3922 vegetative cells and spores in apple juice. LWT - Food Sci. Technol. 2015, 60, 238–245, doi:10.1016/j.lwt.2014.07.029.

43. Chedea, V.S.; Echim, C.; Braicu, C.; Andjelkovic, M.; Verhe, R.; Socaciu, C. Composition in polyphenols and stability of the aqueous grape seed extract from the romanian variety "merlot recas." J. Food Biochem. 2011, 35, 92–108, doi:10.1111/j.1745-4514.2010.00368.x.

44. Iuga, M.; Mironeasa, S. Potential of grape byproducts as functional ingredients in baked goods and pasta. Compr. Rev. Food Sci. Food Saf. 2020, 19, 2473–2505, doi:10.1111/1541-4337.12597.

45. Pop, R.M.; Măgureanu, D.C.; Macovei, Ştefan O.; Chedea, V.S. Introduction to nutraceuticals. In Nutraceuticals: Evidence Based Use; Pop, R.M., Ed.; Nova Science Publishers, Inc.: New York, 2020; pp. 1–39.

46. Chedea, V.S.; Palade, L.M.; Marin, D.E.; Pelmus, R.S.; Habeanu, M.; Rotar, M.C.; Gras, M.A.; Pistol, G.C.; Taranu, I. Intestinal absorption and antioxidant activity of grape pomace polyphenols. Nutrients 2018, 10, doi:10.3390/nu10050588.

47. Chedea, V.S.; Pelmus, R.S.; Lazar, C.; Cecilia, G.; Calin, L.G.; Toma, S.M.; Dragomir, C. Effects of a diet containing dried grape pomace on blood metabolites and milk composition of dairy cows. J. Sci. Food Agric. 2017, 97(8), 2516–2523.

48. Chedea, V.S.; Palade, L.M.; Pelmus, R.S.; Dragomir, C.; Taranu, I.; Chedea, V.S.; Palade, L.M.; Pelmus, R.S.; Dragomir, C.; Taranu, I. Red grape pomace rich in polyphenols diet increases the antioxidant status in key organs—Kidneys, liver, and spleen of piglets. Animals 2019, 9, 149, doi:10.3390/ani9040149.

49. Zhang, X.-H.; Choi, S.-K.; Seo, J.-S. Effect of dietary grape pomace on lipid oxidation and related enzyme activities in rats fed high fat diet. Korean J. Nutr. 2009, 42, 415, doi:10.4163/kjn.2009.42.5.415.

50. Yu, J.; Bansode, R.R.; Smith, I.N.; Hurley, S.L. Impact of grape pomace consumption on the blood lipid profile and liver genes associated with lipid metabolism of young rats. Food Funct. 2017, 8, 2731–2738, doi:10.1039/c7fo00542c.

51. Şehirli, Ö.; Ozel, Y.; Dulundu, E.; Topaloglu, U.; Ercan, F.; Şener, G. Grape seed extract treatment reduces hepatic ischemia-reperfusion injury in rats. Phyther. Res. 2008, 22, 43–48, doi:10.1002/ptr.2256.

52. Rivera, K.; Salas-Pérez, F.; Echeverría, G.; Urquiaga, I.; Dicenta, S.; Pérez, D.; de la Cerda, P.; González, L.; Andia, M.E.; Uribe, S.; et al. Red wine grape pomace attenuates atherosclerosis and myocardial damage and increases survival in association with improved plasma antioxidant activity in a murine model of lethal ischemic heart disease. Nutrients 2019, 11, 2135, doi:10.3390/nu11092135.

53. Shao, Z.H.; Hsu, C.W.; Chang, W.T.; Waypa, G.B.; Li, J.; Li, D.; Li, C.Q.; Anderson, T.; Qin, Y.; Schumacker, P.T.; et al. Cytotoxicity induced by grape seed proanthocyanidins: Role of nitric oxide. Cell Biol. Toxicol. 2006, 22, 149–158, doi:10.1007/s10565-006-0052-x.

54. Yamakoshi, J.; Saito, M.; Kataoka, S.; Kikuchi, M. Safety evaluation of proanthocyanidin-rich extract from grape seeds. Food Chem. Toxicol. 2002, 40, 599–607, doi:10.1016/S0278-6915(02)00006-6.

55. Bentivegna, S.S.; Whitney, K.M. Subchronic 3-month oral toxicity study of grape seed and grape skin extracts. Food Chem. Toxicol. 2002, 40, 1731–1743, doi:10.1016/S0278-6915(02)00155-2.

56. Charradi, K.; Mahmoudi, M.; Bedhiafi, T.; Jebari, K.; El May, M.V.; Limam, F.; Aouani, E. Safety evaluation, anti-oxidative and anti-inflammatory effects of subchronically dietary supplemented high dosing grape seed powder (GSP) to healthy rat. Biomed. Pharmacother. 2018, 107, 534–546, doi:10.1016/j.biopha.2018.08.031.

57. Sivaprakasapillai, B.; Edirisinghe, I.; Randolph, J.; Steinberg, F.; Kappagoda, T. Effect of grape seed extract on blood pressure in subjects with the metabolic syndrome. Metabolism 2009, 58, 1743–1746, doi:10.1016/j.metabol.2009.05.030.

58. Rychen, G.; Aquilina, G.; Azimonti, G.; Bampidis, V.; de Lourdes Bastos, M.; Bories, G.; Cocconcelli, P.S.; Flachowsky, G.; Gropp, J.; Kolar, B.; et al. Safety and efficacy of dry grape extract when used as flavouring in water for drinking for all animal species and categories. EFSA J. 2016, 14, doi:10.2903/j.efsa.2016.4627.

Chapter 3

Bacteriocins: Nisin as an Alternative Source to Chemical Preservatives

Abeer I. M. El-Sayed,[a] Ibrahim M. El-Sayed,[b] and Sameh Awad[b]
[a]*Department of Botany and Microbiology, Faculty of Science, Damanhour University, EL-Beheira, Egypt*
[b]*Department of Dairy Science and Technology, Faculty of Agriculture, Alexandria University, Alexandria, Egypt*
abeeribrahimelsayed@hotmail.com, elsayedibrahim201@yahoo.com, sameh111eg@yahoo.com

Owing to the high nutritional content of animal foods such as meat and dairy products, such foods need a proper preservation method to keep their quality and safety, otherwise these foods will cause economic losses and threaten public health. Biopreservation of foods using bacteriocins or bacteriocin-producing strains can assure the safety of foods by inhibiting the growth of pathogenic bacteria and food spoilage microorganisms.

Many bacterial species belonging to Gram-positive or Gram-negative bacteria have the ability to produce inhibitory substances

Natural Food Preservatives
Edited by Sourish Bhattacharya and Heba Hassan Abd-El Azim Salama

ISBN 978-981-4968-64-5 (Hardcover), 978-1-003-36776-5 (eBook)
www.jennystanford.com

known as bacteriocins. These bacteriocins are inhibitory agents that have a proteinaceous or peptidic nature and display antimicrobial activity against some other strains closely related to the bacteriocin-producing strain but do not harm the producing bacterial cell.

Bacteriocins produced by Gram-positive bacteria such as lactic acid bacteria can be classified into diverse groups and some of them can be used as biopreservatives in the food industry like Nisin produced by *Lactococcus lactis* subsp. *lactis* to prevent spoilage and pathogenic microorganisms.

3.1 Biopreservation of Food

Food preservation and safety are currently the primary concerns of both consumers and food producers. Food preservation is heavily influenced by the product's quality and safety. It is a critical issue to prevent foodborne illnesses, which are both serious and expensive. Food preservation aims to maintain the product's safety and overall quality (organoleptic and nutritional), as well as to minimize waste, eventually expand the product's shelf life, and preserve food materials during transportation [15]. There are many mechanisms for preserving food, which include physical techniques like low-temperature conservation (refrigeration and freezing) and the use of heat treatments (pasteurization, boiling, sterilization, etc.). Chemical preservatives (sodium chloride, sodium nitrite, benzoates, sorbates, etc.) are also used to preserve foods. These classical preservation techniques may pose health risks, alter the sensory properties of food, and result in the loss of several nutritive entities. As a result, biopreservation techniques appear to be more practical for improving food quality and safety. Furthermore, biopreservation has the potential to increase the food shelf life while maintaining hygienic standards and having minimal negative effects on nutritional and organoleptic properties. Currently, there is a growing interest in replacing traditional food preservation methods with innovative technologies that include biological antimicrobial systems such as lactic acid bacteria (LAB) or their metabolites. Food biopreservation is the use of LAB and/or their metabolites to extend the shelf life of foods [35]. Aside from the use of LAB and their metabolites, food biopreservation methods focus on the use of bacteriophages and bacteriophage-encoded enzymes (e.g., endolysins). These applications are critical for eliminating any use of chemical preservatives or harsh thermal processes, both of which can have a negative impact on the safety or quality of food products.

3.2 Methods of Food Biopreservation

3.2.1 Bacteriophages

Bacteriophages, which are true parasites, are classified as virulent or temperate viruses which can only reproduce within bacterial cells and unable to infect or harm human or animal cells [51]. Bacteriophages are the most common microorganisms in foods.

Virulent bacteriophages can lyse the host bacterium and have antimicrobial properties. Bacteriophages could be used effectively in the field of food preservation to avoid pathogen contamination or colonization on food surfaces. These bacteriophages were found to be effective in controlling *Salmonella* spp., *Campylobacter* spp., and *Escherichia coli* in poultry and ruminants. Bacteriophages seemed to be effective at suppressing *Salmonella, Staphylococcus aureus,* and *Listeria monocytogenes* in dairy products [29]. Some bacteriophage-based products have been approved by the US Food and Drug Administration (FDA) and are now available for commercial use. However, more research is required to resolve many of the restrictions of phage applications in food, such as the rise of resistance and intolerance to environmental conditions.

3.2.2 Endolysins

Known as phage lysins which are phage-encoded peptidoglycan hydrolases used by a vast bulk of bacteriophages to enzymatically destroy the host bacterium's peptidoglycan layer. Endolysin treatment is a novel and promising solution to tackle antibiotic-resistant bacteria, which are a common problem in the food industry [5]. Endolysins, also, have a potent antimicrobial property, especially against Gram-positive bacteria. These endolysins have the potential to treat mastitis and prevent biofilms made from *S. aureus* and *Listeria* sp.

3.2.3 Protective Cultures

The fermentation of food is a common example of biological food preservation. It results in the formation of a massive number of bacterial cells as well as the accumulation of favorable metabolites, which may minimize the incidence of food spoilage and pathogenic microorganisms [21]. Biological preservatives that are suitable for use in food must be recognized as a safe, nonpathogenic, nontoxic producer that is also able to persist in food. Starter and protective cultures are the most common biological agents used throughout food biopreservation. Starter cultures are primarily used to initiate fermentation, produce organic acids, and generate flavor and aroma compounds. However, because of their antimicrobial activity, protective cultures are widely used to manage the growth and survival of many pathogenic microorganisms in foods.

3.3 History and Identification of Bacteriocins

Bacteriocins started to catch the attention of scientists from the middle of1920s, exactly when Gratia (1925) discovered the bacteriocin (colicin V), and this was continued by Jacob [26] who named these peptides "bacteriocins." Till now, bacteriocins are identified as extracellular antimicrobial ribosomally synthesized peptides, which can be produced by both Gram-positive and Gram-negative bacteria and act against closely related bacterial species as stated by Daba and Elkhateeb [10].

Owing to the well-known applications of some LAB strains, which are Gram positive, in food industries for many years ago, the American Food and Drug Administration (FDA) classified a lot of LAB products and bacteriocins that can inhibit a wide range of microorganisms including many food spoilage and foodborne pathogens as those relating to the genera of *Salmonella* and *Listeria*; to be generally recognized as safe (GRAS) as a human food ingredient [2]. Owing to the spread of serious foodborne illnesses and consumer demand for limiting chemical additives, bacteriocins gain their chance to attract interest as safe and natural antimicrobial agents for food preservation [45]. It is well known that bacteriocins produced by LAB are thermostable at high thermal stress, colorless, odorless, tasteless, stay active over a wide pH range, and easily degraded by the action of proteolytic enzymes. Additionally, some bacteriocins have high specificity against many bacterial strains, and all these favorable properties made bacteriocins used in an ideal way for industrial food preservation combating pathogenic and spoilage microorganisms, thus increasing their shelf life [10].

3.4 Classification of Bacteriocins

For a clear bacteriocin classification, many features must be taken into consideration, such as size, homology, structure, and posttranslational modifications. However, these classifications are frequently revised due to appearance of novel and variable bacteriocins with new properties [9]. Bacteriocins are categorized into various classes based on many criteria as producer organisms, molecular sizes, physical properties, chemical structures, and mode of actions. Initially, Klaenhammer [31] divided bacteriocins into the following four classes:

(i) Lantibiotics were put under class I and characterized by being thermostable with very low molecular weight (<5 kDa) and the presence of lanthionine and its derivatives. Nisin can be described as an important member of this class.

(ii) Class II comprises small thermostable peptides without lanthionine derivatives and up to 10 kDa molecular weight. It was also grouped into three subclasses as IIa (pediocin and enterocin), IIb (lactocin G), and IIc (lactocin B).

(iii) Class III contains bacteriocins with high molecular weight (>30 kDa) and thermolabile peptides like helveticin J.

(iv) Class IV includes large peptides combined with carbohydrates or lipids [3, 31]. However, Cleveland et al. [7] proposed a contradiction regarding the large peptides of class IV and described them as artifacts of partial purification and not considered a new class of bacteriocins.

Four years later, Cotter et al. [9] suggested a new bacteriocin classification and showed only two classes in this concept: Class I (lantibiotics) and class II (other peptides without lanthionine). With excluding the thermolabile and high molecular weight peptides from these two classes with a separate group for bacteriolysins and recommended that class IV in the initial classification must be extinguished.

One year later, Drider et al. [17] finally grouped bacteriocins into three main classes according to their genetic and biochemical features. Table 3.1 shows the basic properties of this classification [2, 17].

Table 3.1 Classification of bacteriocins

Classification	Features	Subcategories	Examples
Class I (Lantibiotics)	Lanthionine or peptides containing β-lanthionine	Type A (linear molecules) Type B (globular molecules)	Nisin, subtilin, epidermin, mersacidin
Class II	Heterogeneous class of small thermostable peptides	Subclass IIa (antilisterial-pediocin bacteriocins type) Subclass IIb (composed of two peptides) Subclass IIc (other bacteriocins)	Pediocin, enterocin, sakacin, plantaricin, lactacin F, lactococcin
Class III	Large thermolabile peptides		Helveticin J, millericin B

3.4.1 Class I

Containing bacteriocins, called lantibiotics, described as being small (19–38 amino acid residues) and thermostable peptides. Nisin is the famous representative for this group with a structure of two functionally different domains: i) the N-terminal domain that comprises two rings forming a high affinity to pyrophosphate groups of lipid II; ii) the C-terminal domain, which is critical for pore formation. Many strains of *L. lactis* subsp. *lactis* are natural producers for this bacteriocin type consisting of 34 amino acid residues. Nisin A and nisin Z have the same antimicrobial activity and molecular structure, except for only one amino acid. Additionally, another new nisin variant isolated from *Streptococcus uberis*, called nisin U, has 78% similarity with nisin A [2, 50].

Nisin has a broad antimicrobial spectrum range on many pathogens and LAB species concerning *L. monocytogenes*, *S. aureus*, and *Bacillus cereus*. This makes nisin a critical tool for many technological applications.

Nisin's action finally ends with cell death through lipid II (docking molecule) by pore formation mechanism, causing some changes in the membrane permeability of the target cell and outflow of essential compounds (K^+ ion, ATP, and amino acids) through these pores (Fig. 3.1A) [23, 50].

3.4.2 Class II

Bacteriocins in this class are structurally simpler than lantibiotics as they do not have posttranslational modifications in the peptide chain such as lanthionine or β-lanthionine. These bacteriocins include an amphiphilic helical form of thermostable small (<10 kDa) peptides. Several class II bacteriocins, including pediocin PA-1, sakacin P, and lactococcin A, bind to the receptor Man-PTS [47], a transport system that couples sugar import and phosphorylation (Fig. 3.1A and B). This conformation structure makes the insertion or enrollment of class II bacteriocin in the cytoplasmic membrane of the target cell easy, resulting in membrane depolarization and cell death. Moreover, class II bacteriocins can be divided into three subclasses, namely, subclass II-A, subclass II-B, and subclass II-C [2, 50].

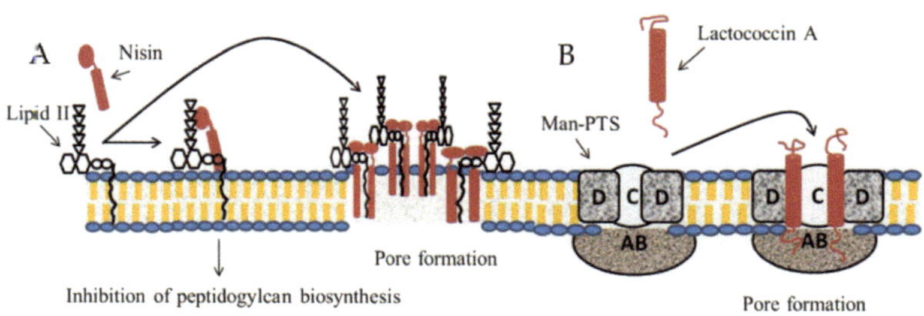

Figure 3.1 Mechanisms of action of nisin and lactococcin A. (A) By targeting lipid II, nisin can inhibit peptidoglycan biosynthesis and form pores in the bacterial membrane. (B) Lactococcin A uses the mannose phosphotransferase system (Man-PTS) as a receptor, leading to the uncontrolled opening of this receptor and thereby forming a pore in the bacterial membrane [50].

3.4.2.1 Subclass II-A

Bacteriocins of this subclass are recognized for being with high antilisterial activity of molecular structure including 37–48 amino acid residues. The N terminal is in the form of a pleated sheet and the C terminal has one or two α-helices. Regarding the mode of action for this bacteriocins category, bacteriocin falls into the cell membrane of the target cell by the C terminus and starts a pore formation with consequent dissipation of proton motive force causing high ATP consumption and ending with cell death. Pediocin, sakacin, and enterocin are the well-known representatives of subclass II-A [2].

3.4.2.2 Subclass II-B

These bacteriocins are heterodimeric, consisting of two peptides. Bacteriocin members in this subclass have three criteria: (i) For full antimicrobial activity, both peptides are needed, while the individual peptides give little or sometimes no activity, (ii) one immunity protein is enough to get immunity, and (iii) their genetic organization system comprises two sequential bacteriocin structural genes encoding the single peptides plus a single immunity gene. Lactococcin G is the first well-known bacteriocin related to this group with an antimicrobial activity that depends on both α- and β-peptides. Plantaricin and lactacin F are essential representative members of the two-peptide bacteriocins. Their mechanism of action involves membrane potential dissipation by lowering the intracellular ATP concentration level [2, 6].

3.4.2.3 Subclass II-C

These bacteriocins are unique in their structure as they are formed in a round circular model with a covalent bond between C and N terminals forming head to tail cyclic shape of the peptides. The famous representative member of this subclass is AS-48 from *E. faecalis* with an action mechanism permeabilizing the cytoplasmic membrane of target cells, leading to a dissipation of the proton motive force and finally cell death [2].

3.4.3 Class III

This class is composed of large thermolabile bacteriocins, with more than 30 kDa molecular weight. Class III bacteriocins have complex protein structure and activity providing a completely different mode of action from other bacteriocins, as they can induce lysis of the target microorganism's cell wall. The N-terminal portion of the molecule acts as an endopeptidase while the C-terminal portion recognizes the target cell [2].

3.5 Bacterial Origin of Bacteriocins

3.5.1 Bacteriocins from Gram-Negative Bacteria

Initially, bacteriocins arise from Gram-negative bacteria as stated by Gratia in 1925; *E. coli* can produce colicin, which was the first member described in the bacteriocin family and was identified as an antimicrobial protein. In further investigations, it was detected that not only *E. coli* which can produce bacteriocins, but also many Gram-negative species have this ability in bacteriocin production as klebicins of *Klebsiella pneumonia*, marcescins of *Serratia marcescens*, alveicins of *Hafnia alvei*, cloacins of *Enterobacter cloacae*, and pyocins of Pseudomonas are listed from the most famous examples of bacteriocins produced from other Gram-negative bacterial species [23]. These bacteriocins are characterized by being heat-labile and large peptides, with the exclusion of microcin V of *E. coli* which is heat stable and consists of a few peptides.

The main disadvantage of Gram-negative bacteriocins was noticed with their narrow antimicrobial activity spectrum limiting their industrial use. This undesired feature calls attention for more suitable Gram-positive bacterial bacteriocins [2, 7, 44, 45].

3.5.2 Bacteriocins from Gram-Positive Bacteria

A wide variety of bacteriocins can be found under Gram-positive bacterial producers. Characterized by being with broader inhibitory spectra with non-toxicity property toward eukaryotic cells, these bacteriocins can find lots of industrial and medical uses. LAB group which converts 50% as a minimum, from the carbon of sugars to two lactic acid isomers, through the fermentation pathway process with producing lactic acid. LAB are Gram-positive, heterogeneous, nonmotile, non-spore-forming, rod- or cocci-shaped microbial group that contains 12 bacterial species, including *Lactococcus, Streptococcus, Pediococcus, Lactobacillus, Leuconostoc, Enterococcus, Carnobacterium, Oenococcus, Aerococcus, Tetragenococcus, Vagococcus*, and *Weissella*, with a variety of physiological characters and important metabolic products with safe nature (e.g., organic acids, diacetyl, acetoin, hydrogen peroxide, reuterin, reutericyclin, antifungal peptides) and in addition their ability to produce inhibitory peptide molecules as the valuable bacteriocins given in Table 3.2 [2, 6, 7, 23, 45].

Table 3.2 Some important bacteriocins produced from LAB

Bacteriocin	**Producer organism**
Nisin A	*L. lactis*
Nisin Z	*L. lactis*
Nisin U	*S. uberis*
Lactococcin DR	*L. lactis* ADRIA 85L030
Lacticin 481	*L. lactis* CNRZ481
Lacticin 3147 (LtnA1 and LtnA2)	*L. lactis* DPC3147
Estreptococcin A-FF22	*S. pyogenes* FF22
Salivaricin A	*S. salivarius* 20P3
Cytolysin (CylL1 and CylL2)	*E. faecalis*
Carnocin U149	*C. piscicola U149*
Lactocin S	*L. sakei* L45
Pediocin PA1	*P. acidilactici* PAC-1.0
Pediocin AcH	*P. acidilactici H*
Leucocin A-UAL187	*L. gelidum UAL187*
Mesentericin Y105	*L. mesenteroides Y105*

Source: [6, 23]

3.6 Biosynthesis of Bacteriocins

Bacteriocins are considered biological products of microorganisms that have certain and specific features to be distinguished from other biologically synthesized molecules. These features create the biological properties of bacteriocins including synthesis, mode of action, immunity, and resistance mechanisms.

All bacteriocins are of proteinaceous nature, as they are synthesized by ribosomes found in the producer microorganisms. The genetic code occasionally placed on plasmids or the main chromosome or mobile elements as transposons can supply the required information [6, 45].

The bacteriocin gene clusters genetically organized as a functional operon consists of the structural gene, the gene codifying immunity protein, the genes responsible for synthesizing and transporting the bacteriocin, and, in some cases, the genes that can codify many enzymes for posttranscriptional modifications. Moreover, bacteriocin operons also comprise the pre-induction factor (IF) coding gene, the histidine protein kinase (HPK) gene, and the gene(s) codifying the response regulator (RR) [7].

At the beginning of the bacteriocin production, the structural gene is activated producing biologically inactive precursors or pre-propeptides identified as pre-probacteriocins. Then, other related parts of the bacteriocin operon become operational sequentially releasing the mature bacteriocins with antimicrobial activity [6].

Numerous studies demonstrate that the competition between microflora present in nutrient-limited conditions plays a critical role in controlling the production of bacteriocin. IF (induction factor), a class of bacteriocin-like peptides with lengths of 19 to 26 amino acids and low molecular weights and cationic nature, may also play a significant role in the regulation mechanism [2, 6, 45].

External IFs can regulate the expression of bacteriocin genes. Production of bacteriocin depends on the temperature, pH, and other environmental circumstances or it can be constitutive. Bacteriocin production may be controlled by a "quorum-sensing" mechanism in some LAB. The quorum-sensing system limits the excess of some important biological processes in bacteria such as natural genetic transformation, virulence, and sporulation. In Gram-positive bacteria, peptide pheromones or IF mediate the quorum sensing. Signal transduction pathway is composed of three components in which IFs are the first one that can bind and activate the correspondent histidine protein kinase to phosphorylate the response regulator, which stimulates the target genes transcription, frequently, by binding to specific imperfect direct repeats present in many bacteriocin genes cluster. Bacteriocin producer strain or other strains belonging to the same or other species or genera can produce IF for a given bacteriocin operon. When the bacteriocin producer itself produces the inducer peptide, the autoregulation of biosynthesis could be achieved [16].

Two models were provided to explain the induction of bacteriocin and the significant involvement of IFs. The quorum sensing model was the first one, which illustrates that the IF is constitutively formed and amassed in low concentrations within bacterial growth. When IF concentration reaches the threshold for IF auto-induction level, the bacteriocin genes are then induced. Thus, this model builds on a control mechanism depending on the cell density of the strains.

In the second model, which is an alternative for the quorum sensing discussed above, the concentration of the IF never reaches its threshold by itself and needs unknown environmental signals or some modification in environmental conditions such as variations in nutrient levels or physicochemical growth conditions [6].

Moreover, many recent studies listed that three components build the regulating system in many cases: an inducing peptide (or pheromone-activating factor), the transmembrane histidine kinase (pheromone receptor), and a response regulator. Apart from these, it is also known that lantibiotics (nisin and plantaricin) production is regulated and directly controlled by the bacteriocin itself, acting as pheromone inducing their production at high levels [2].

A critical difference between bacteriocins and antibiotics is the immunity mechanism of the strain that produces bacteriocin to its own product. As the producer strain makes self-protection for the toxicity of its own bacteriocins via a certain mechanism relying on many bacteriocin-specific immunity proteins which are completely different from the bacteriocin resistance mechanism, that is explained by the spontaneous and/or induced mutations help in a new resistant strains formation with many changes in both cell wall and cell membrane as electrical potential changes, fluidity, and membrane lipid composition changes, cell wall thickness is also noticed and alterations in bacteriocin receptors [2, 6, 7].

3.7 Mode of Bacteriocins' Action Differs from Antibiotics

Commonly bacteriocins are synthesized as inactive prepeptides having an N-terminal sequence guide [37]. During the exponential growth phase, these precursors are transported to the cell surface and enzymatically converted into their active forms. Most of the bacteriocins given by LAB are preferentially active against closely related species, but Nisin has an antimicrobial effect against a wide range of Gram-positive bacteria, including *Bacillus* sp. and *Clostridium botulinum*. Bacteriocins can inhibit antibiotic-resistant microorganisms by exerting a bactericidal effect at the membrane level [56]. Bacteriocins' thermal and pH stability makes them a particularly attractive candidate for use as a food preservative. Bacteriocin-producing genetic determinants are typically found on plasmids, which can be moved from donor to recipient cells, improving the characteristics of the bacteriocin-producing strain. The use of bacteriocins as a food preservative has several advantages, including [1] minimizing food poisoning, cross-contamination, and the use of chemical preservatives and [2] improving food product quality and shelf life [19]. Bacteriocins can be used in a variety of ways in foods, including the use of producer strains and the addition of free or encapsulated bacteriocins [19]. Bacteriocin action is primarily attributed to the formation of a pore in the bacterial cell membrane, which results in the efflux of small molecules such as amino acids, potassium ions, and ATP [36]. Other mechanisms have been proposed to explain how bacteriocins act on bacterial cells. Bacteriocins could activate cell wall acting enzymes (amidases) by releasing the enzymes from their inhibitors (teichoic and teichuronic acids), resulting in bacterial autolysis. Bacteriocins may also act on specific intracellular targets, such as inducing proteolytic enzyme production, causing metabolic disruption, and impairing respiratory function. This could illustrate the formation of lysis vesicles in cells of *L. innocua* and *L. casei* subsp. *casei* treated with nisin Z, as previously reported by Benech et al. [3].

Three methods could be used to apply LAB-producing bacteriocins and their metabolites to foods. The first is the direct application of LAB-producing bacteriocins, which incorporate the bacteriocin into foods during processing. The second method is to directly add purified or crude bacteriocins to the food product, and the third method is to use a previously fermented product from a bacteriocin-producing strain [35].

For class II, the system regulating bacteriocin production is formed of three components: an inducing peptide (or pheromone-activating factor), the transmembrane histidine kinase (pheromone receptor), and a response regulator. The inducing peptide inducer is synthesized at low levels in the ribosome as a prepeptide, which is cleaved and secreted in the surrounding environment by the carrier system. When reaching a threshold concentration by this compound, the transmembrane histidine kinase is activated, which leads to autophosphorylation of the histidine residue and, consequently, phosphate is transferred to a response regulator protein. The phosphorylated regulator activates the transcription of the bacteriocin in addition to the elements that form the regulatory system. Production regulation of lantibiotics such as nisin and subtilin is made by the bacteriocin itself, which acts as a pheromone inducing their production at high levels [33].

Bacteriocins' lethal action starts by entering the target cell by the interaction with specific molecules on the cell surface acting as receptors. Mannose-phosphotransferase systems (Man-PTS) and lipid II have been identified as receptors. Class IIa bacteriocins target a subgroup of Man-PTS on sensitive cells of genera such as *Listeria*, *Enterococcus*, and *Lactobacillus* [30]. The artificial role is played by the N-terminal portion of class II bacteriocins in electrostatic binding to the membrane surface, while for determining the target specificity, the C-terminal region is crucial [24].

In cell wall synthesis, lipid II has an important role, serving as an anchoring receptor for the vancomycin/teicoplanin family of glycopeptides, mannopeptimycins, ramoplanins, katanosins/plusbacins, and lantibiotics. While vancomycin binds the D-Ala-D-Ala peptide side chain of lipid II, the lantibiotic nisin binds the highly conserved sugar-pyrophosphate moiety. Vancomycin-resistant strains thus remain sensitive to nisin [25].

Insertion of bacteriocin into the cytoplasmic membrane to form ion-permeable channels then large pores; are due to binding to the membrane surface (through specific or nonspecific receptors). Death of microbial cell may be a result of efflux of metabolites, dissipation of vital ion gradients, leakage of the cell contents, random degradation of cellular DNA, low protein synthesis through specific cleavage of 16S rRNA, or lysis of the cell due to inhibition of peptidoglycan synthesis [58]. As an example, colicin E2 kills the target cell after translocation into the periplasmic space with nonspecific cleavage of DNA, colicin E3 can kill via ribosome inactivation, while colicin M reduces murein biosynthesis via hydrolyzing the pyrophosphate linkage joining bactoprenol and murein precursor [32]. The intracellular modes of action for antibiotics and bacteriocins are different. Another example, macrolides are from the important classes of antibiotics that can block protein synthesis by binding to the nascent peptide exit tunnel of the ribosome. RNase-type colicins lower protein synthesis through cleavage of a specific site near the 30 end of 16S rRNA [28].

3.7.1 Self-Immunity Mechanisms

Certain specific protection mechanisms called self-immunity mechanisms must be employed to prevent the lethal effect of bacteriocin production on the producer strains [50]. The antibacterial activity of bacteriocins produced by Gram-positive bacteria is performed by targeting the membrane of the bacterial cell especially lipid II or Man-PTS; in order to protect these components in the producer strain, the ABC transport system and the self-immunity proteins are involved. So, the cell membrane bound bacteriocins produced by Gram-positive bacteria must be rapidly removed from the membrane of the producer strain; thus, the producer strain is protected from the lethal effect of its own bacteriocin. It must be taken into consideration that the ABC transporter involved in this self-immunity mechanism is not the same ABC transporter that is used for bacteriocin translocated outside the cell, and also this type of transporter is encoded by different genes [50].

A remarkable example of that is given by the mersacidin produced by *B. amyloliquefaciens* (Fig. 3.2A) in which Mrs A "the mersacidin precursor" is modified by Mrs M, forming a tetracyclic structure model, followed by being processed and exported as a mature form of the mersacidin by Mrs T.

Another kind of ABC transporter coded by three genes operon (mrs E, mrs F, and mrs G) is involved in the self-immunity action to ensure a strong protection for the producer strains. Finally, the immunity mechanism for *B. amyloliquefaciens* producing mersacidin is done by the multifunction of a cognate immunity protein that binds to the molecules of the bacteriocin found on the cell membrane with the action of the multicomponent ATP-binding cassette (ABC) transporter that can take the bacteriocin out of the cell [50].

Moreover, the transmembrane proteins share in the self-immunity mechanism can be encountered for some bacteriocins produced by Gram-positive bacteria. The nisin producer strain can be protected against the lethal effect of its own bacteriocin by cooperation with the ABC transporter which can bind the nisin and prevent the death of the producer strain.

The self-immunity of Man-PTS-targeting BGPB (e.g., lactococcin A) is mediated by specialized proteins that tightly bind the bacteriocin, preventing it from affecting the function of the Man-PTS of the producer strain (Fig. 3.2B).

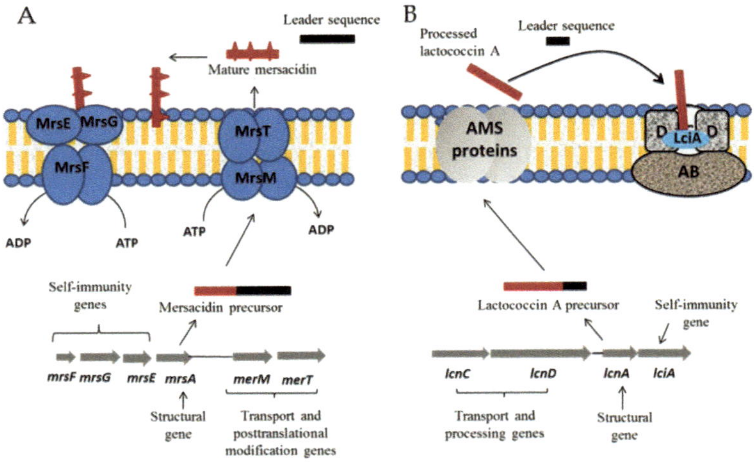

Figure 3.2 Self-immunity and transport mechanisms of mersacidin (A) and lactococcin A (B) that target lipid II and Man-PTS, respectively. Proteins employed for self-immunity and transport of these bacteriocins are distinct and encoded by different genetic elements [50].

In the case of microcins and bacteriocins, till now the immune system of the producer strains remains unknown, while one of the hypotheses explained that the lethal peptide of the microcin can be removed outside the producer strain by special proteins appear to be an efflux pump as a mean of the immune system [50].

A specific enzyme called "serine peptidase" can deactivate intact microcin C through breaking the amide bond that links peptidyl moieties with the nucleotide part. Recent studies have reported that the same enzyme plays an important role in the sensitive strains for conferring their resistance to microcin C as illustrated in Fig. 3.3. Microcins categorized under class II bacteriocins showed a self-immune system consisting of two or three transmembrane proteins which are able to hardly bind with the microcin, thus preventing its reaction with the membrane.

Regarding the self-immune mechanism of the colicin group, some proteins make a hard bond with the catalytic C terminal of the colicins resulting in blocking the active site of the enzyme or the substrate binding site. An inner membrane channel in the producer strain is not formed due to the action of immunity proteins of pore-forming colicins [50].

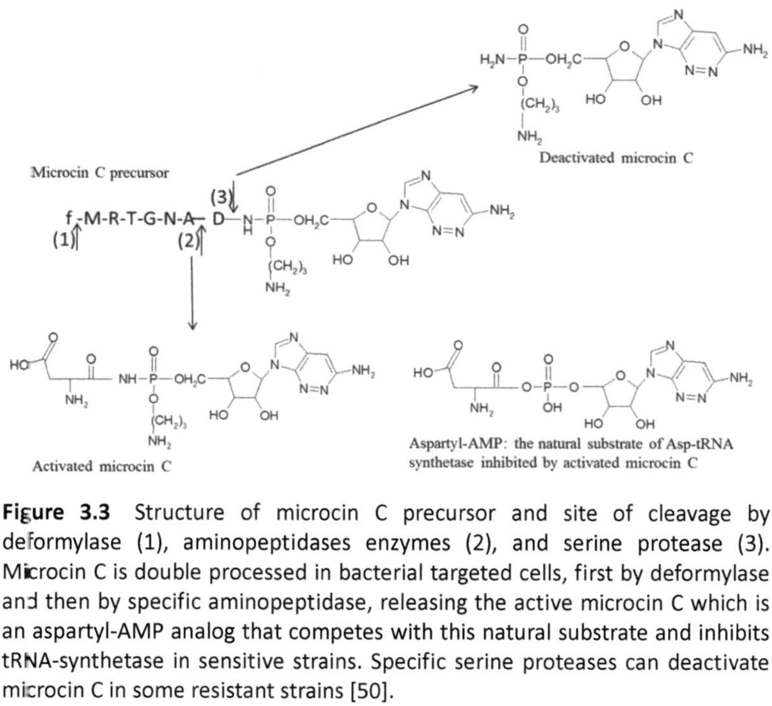

Figure 3.3 Structure of microcin C precursor and site of cleavage by deformylase (1), aminopeptidases enzymes (2), and serine protease (3). Microcin C is double processed in bacterial targeted cells, first by deformylase and then by specific aminopeptidase, releasing the active microcin C which is an aspartyl-AMP analog that competes with this natural substrate and inhibits tRNA-synthetase in sensitive strains. Specific serine proteases can deactivate microcin C in some resistant strains [50].

3.8 Nisin

Some *L. lactis* strains produced this bacteriocin. The name (Nisin) is derived from the term "N-inhibitory substance" (NIS) plus the traditional classical suffix "in" [7, 11]. In (1928), Rogers [47] became the first scientist to illustrate nisin which can reduce *L. bulgaricus* growth. Additionally, it is well known that the dry state of nisin is more stable under controlled recommended storage conditions. However, due to the acidic nature of the nisin molecule, its thermostability hardly relies on the pH of solutions or food systems [55]. Delves-Broughton [13] found that nisin is totally stable at pH 2.0 and can be stored at temperatures between 2°C and 7°C for a long time, but deactivation happens at a pH above 7.0 even at room temperature. Information on kinetics and bacteriocins' thermal inactivation is critical to enable their proper function as biopreservatives in the high-temperature processes of the food industry [12].

Nisin is an antibacterial thioether-bridged pentacyclic peptide formed from 34 residues related to a cationic peptide group of antimicrobials which is called Type A (I) lantibiotics. Initially, nisin was identified within fermented milk cultures; nowadays, it is a common natural and safe food biopreservative in many food products, such as processed cheese, meat dairy desserts, fermented beverages, milk, and canned foods [40].

European Union (E234), World Health Organization (WHO), and the US Food and Drug Administration (FDA) have approved nisin usage. Now nisin is licensed in over 50 countries [42], and it has had a high safety profile over the past 40 years of usage with its strong antimicrobial action against a lot of Gram-positive microorganisms of food spoilage and pathogenic bacteria including *S. aureus* and *L. monocytogernes*. It can be effective at nanomolar concentrations relying on the target strain.

Regarding the nisin structure, it is a ribosomally synthesized prepeptide, biologically inactive molecule containing a C-terminal prepeptide domain, following a variety of posttranslational modification reactions, and cleaved at the N-terminal leader sequence to obtain the mature antimicrobial compound with unusual amino acids including dehydroalanine and dehydrobutyrine, made from serine dehydration plus the threonine residues, respectively. These residues are coupled with the thiol group of cysteines obtaining lanthionine and β-methyl lanthionine, that is later enzymatically modified [8, 42].

3.8.1 Nisin as a Biopreservative

Biopreservation is defined as the use of natural microflora in the form of protective cultures and/or their metabolites as antimicrobial agents to improve food safety and storage time. Bacteriocins derived from lactic acid bacteria (LAB) are widely recognized as a natural method of food biopreservation. Bacteriocins have been used in food products using three different strategies: purified addition, in situ production by bacteriocinogenic bacteria, and binding of bacteriocins to polymeric packaging [20]. Bacteriocins such as purified or semi-purified nisin and enterocin AS-48 can be used as biopreservatives in food to extend shelf life and reduce the usage of classical preservatives such as nitrates with negative influence on human health. Another feasible approach is to introduce the bacteriocinogenic strains (particularly LAB) as a starter culture meaningfully into fermented sausage or dairy products without purification to improve food quality and safety [14]. The development of antimicrobial packaging films is a viable alternative for preserving a complicated food environment [39]. Because of their natural and safe properties, bacteriocins are the most fascinating antimicrobial agents in this regard [7]. Nowadays, food preservation is frequently accomplished through the use of authorized chemical preservatives, the most common of which are benzoic and sorbic acids, as well as their sodium, potassium, and calcium salts. Sodium salts, in particular, are widely used in the production of cheese. In rare circumstances, these chemical preservatives may be harmful to consumers' health. In humans, for example, allergic reactions to benzoates have been reported, including urticaria, non-immunological contact urticaria, and asthma. At present, there is a greater consumer demand for natural and minimally processed foods, owing to increased consumption. The bacteriocin nisin, produced by *L. lactis* strains, is generally regarded as safe (GRAS) and has been approved for use as a food additive by the World Health Organization. Nisin has gained a lot of attention in the food industry because of its strong inhibitory effect on a wide range of Gram-positive organisms, including spoilage and pathogenic bacteria like *S. aureus*, *L. monocytogenes*, and *C. botulinum*. Furthermore, nisin is the only bacteriocin approved for widespread use in the food industry, and numerous recent studies have demonstrated the successful use of nisin against pathogenic microbes in a variety of settings as kinds of cheeses (Camembert, Manchego, Cheddar, Vidiago, Gouda, Cottage cheeses), demonstrating 1–3.5 log of viability loss for many microbial pathogens [57].

According to the studies, the use of bacteriocins with these properties in food industries can extend the shelf life of foods, provide better security throughout temperature misuse conditions, reduce the risk of transmission of foodborne pathogens through the food chain, reduce economic losses due to food spoilage, reduce the use of chemical preservatives, and allow the application of bacteriocins [23]. Table 3.3 was created to summarize the potential bacteriocin applications in the food industry while also detailing these applications [20].

Table 3.3 Bacteriocins in food applications

Potential application	Target groups	Application details
Milk and dairy products	Raw material	– Inhibition of microbial load in raw milk – Stop the activity of mesophilic bacteria in milk
	Fermented products	– Stop gas formation by *C. tyrobutricum* on semihard and hard cheeses types – Inactivation of pathogenic and toxicogenic bacteria (*L. monocytogenes, B. cereus, S. aureus*) in cheese and on the cheese surface – Reduce the activity of mesophilic bacteria and endospore formers in cheese – Regulate the over acidification in yogurt and other fermented products – Bacteriocin-producer strains are used as starter cultures to decrease pathogenic and spoilage bacteria in cheese and other fermented milk products – Bacteriocin-producer strains cultures to inhibit adventitious non-starter LAB microflora in cheese – Cheese ripening is accelerated through the increased release of bacterial intracellular enzymes
	Processed products	– Endospore formers (*C. botulinum*) are inhibited in processed cheese and other processed dairy products – *L. monocytogenes* inhibition in dairy products following post-process contamination

Potential application	Target groups	Application details
Meat and poultry products	Raw materials	– Decontamination of raw meat (e.g., by spray washing) – Inactivation of pathogenic and spoilage bacteria and decrease in shelf life extension of vacuum-packaged raw meat – Inhibition of *L. monocytogenes* on raw meat surfaces and in minced meat
	Cooked meat products	– Reduce the microbial load of spoilage LAB – Shelf life extension of vacuum-packaged sliced meat products – Using bacteriocin-producing LAB as protective cultures against *L. monocytogenes* and spoilage bacteria in vacuum-packaged meats
	Egg and derivatives	– Shelf life elongation – The decreased intensity of thermal treatments – High inactivation of pathogenic bacteria
	Fermented meat products	– Inhibition of spoilage LAB – Inhibition of foodborne pathogens (*Salmonella, L. monocytogenes, S. aureus*) – Use of bacteriocin-producing LAB (mainly, *L. sakei* strains) as starters to restrain *L. monocytogenes* and spoilage bacteria – Enhancement of the predominance of starters during fermentation
Vegetable-based foods and drinks	Raw fruits and vegetables	– Reduction of suppression of *L. monocytogenes* in raw vegetables (sprouts and others) – Reduction of suppression of *L. monocytogenes* and *Salmonella* in fresh-cut produce
	Cooked and pasteurized foods	– Regulate endospore formers in pasteurized foods – Control of *B. cereus* in rice-based foods

(*Continued*)

Table 3.3 (*Continued*)

Potential application	Target groups	Application details
	Canned foods	– Endospore-forming bacteria, both aciduric and nonaciduric, are controlled in canned vegetables – Prevention of spoilage by *C. tyrobutyricum* in canned fruit pulp
	Fruit juices and drinks	– Spoilage by *Alicyclobacillus* in fruit juices and drinks is prevented – Reduction of suppression of *E. coli* O157:H7 and *S. typhimurium* in fruit juices – Inhibition of *L. monocytogenes* and *S. aureus* in soy milk
	Fermented drinks	– Reduction of beer spoilage bacterial numbers – Inhibition of wine spoilage bacteria and control of wine malolactic fermentation – Reduce the amount of added sulfur dioxide in the wine industry – Inhibition of cider spoilage bacteria
	Fermented vegetables	– Improvement of sauerkraut fermentation by using a paired starter culture – Control of kimchi over-ripening by supplying nisin to partially inhibit *Lactobacillus* sp. – Inactivation of *B. subtilis* in rice by introducing a nisin-producer *L. lactis* subsp. *lactis* as a starter – Control of *L. monocytogenes* in kimchi by a bacteriocinogenic strain of *P. acidilactici* – Improvement of Spanish-style fermentation of table olives by a bacteriocinogenic starter culture of *L. plantarum* producing plantaricins S and T
	Other fermented vegetable foods	– Sourdough fermentation was improved – Inhibition of rope-forming bacilli in bread – Decreased the number of foodborne pathogens in traditional cereal-fermented foods

Potential application	Target groups	Application details
Fish and seafood products	General trials	– Inhibition of *L. monocytogenes* by adding purified sakacin P – Mixing of nisin and pediocin combined with CO_2 atmosphere packaging to control *L. monocytogenes* in smoked salmon – Introducing of nisin combined with sodium lactate into smoked rainbow trout to regulate *L. monocytogenes* – Mixed treatments of nisin and/or moderate heat treatments/chemical preservatives for inhibition of *L. monocytogenes* in sturgeon caviar and lobster – Supplying of bavaricin A or nisin Z for shelf life extension of brined shrimp – Bacteriocinogenic *Carnobacterium* strains were used as protective cultures against *L. monocytogenes* in cold-smoked salmon – Use of bacteriocinogenic *Lactobacillus* strains as protective cultures against *L. monocytogenes* in cold-smoked salmon

3.8.2 Applications of Nisin for Fermented Dairy Products

Nisin has been shown to be effective in the microbial control of a variety of dairy products, and its use in the production of low pH cheese can be widely evaluated. The use of such nisin-producing and nisin-resistant starter cultures has proven to be an effective way of incorporating and maintaining this bacteriocin throughout the cheese-making process in order to minimize foodborne pathogenic and spoilage bacteria. *L.c. lactis subsp. lactis TAB50* and its lactase-negative proteinase-negative mutant strain TAB50-M4 were measured and named as useful starter cultures or by-products in semihard cheese made from raw or pasteurized milk, providing protection against contamination of milk or curd by *S. aureus* [46]. Even then, environmental and processing factors such as pH and water activity that are required to enhance nisin production by transconjugal or natural producers should be described for the future application of this technique. In Cheddar cheese, strains of *Lc. lactis ssp. cremoris* and *Lc. lactis* grown in pH-controlled reconstituted skim milk (RSM) produced 2–5 times more nisin than mass cultures incubated in RSM, though nisin production varied depending on the producer bacterial strains and their use in making Cheddar-type cheese, 400 IU nisin g/L was sufficient to avoid the spoilage by *C. sporogenes* for more than 90 days in cheese spreads.

Using a nisin-producing bacterial strain to regulate *L. monocytogenes* in Feta and Camembert cheeses yielded varying results, ranging from partial to non-inhibitory effects. When trying to mix nisin powder with milk for the cheese manufacturing process without adding a starter culture, the microbial load is reduced while the nisin concentration remains unchanged after pasteurization. The addition of nisin at 100 or 500 mg/kg directly reduces plate and anaerobic spore counts in processed cheese after 3 months of storage at 5°C or 21°C, as well as the viability of *B. stearothermophilus* and *B. cereus*. and *B. subtilis* was suppressed by adding 5 mg/kg nisin.

3.8.3 Nisin in Active Packaging

Nisin is an extremely surface-active peptide that can bind to a variety of compounds such as fatty acids of phospholipids; this item makes it appropriate for adsorption to solid surfaces and killing bacterial cells that adhere as a matter of fact. As a result, nisin adsorption may reflect a promising step forward in the development of the active packaging process; additionally, the traditional protective mechanism of packaging is supported by nisin's antimicrobial action. Besides that, the active action of the bacteriocin could be enhanced by regulating its migration into the packaged media, allowing its antimicrobial activity to be preserved. To reach this aim, packaging materials made of polymers (e.g., cellulose-based packaging or polypropylene) may be integrated or coated with the bacteriocin. Mauriello [39] successfully demonstrated a low-density polyethylene film coated with nisin bacteriocin for the suppression of the indicator strain *Micrococcus luteus* during the milk storage phase. Although the activity and release of nisin from the packaging film depend on pH and temperature, the antimicrobial packaging reduced microbial viability and maximum growth levels in raw, pasteurized, and UHT milk. Because nisin's stability and solubility decreased from the optimal pH 2.0 to 6.0, a lower pH and higher temperature are preferred for bacteriocin transfer from the packaging film. Cross-linked hydroxypropyl methylcellulose (HPMC) films containing nisin were also found to be effective against *M. luteus* [49]. Cellulose-based packaging process that the nisin is implicated in was assessed as a fill interleaved between slices of Cheddar cheese under a modified atmosphere [48, 48]. Under such conditions, the population of *L. innocua* and *S. aureus* dropped massively in the week one under refrigeration conditions. Consequently, the shelf life of Cheddar cheese was significantly extended.

3.8.4 Nisin Accomplished with Thermal Treatments

Milk is commonly heated to keep it stable during storage and to ensure consumer microbiological safety. Nisin is known to influence the thermal resistance of microorganisms, which is considered an impediment. The use of nisin in heat treatments has increased the shelf life of milk, even in low-refrigeration situations, allowing relatively mild treatments to replace the current thermal treatments and thus improve sensory quality. Rao et al. [43] demonstrated that when 500 IU mg/L nisin was added to an acidified concentrated buffalo milk model system, *B. stearothermophilus* spores could be highly sterilized at 121°C for 5 min. In a comprehensive study, milk containing 40 IU nisin mL/L treated at 72°C for 15 s had a 7-day retention quality increase when compared to the control, as well as a considerably lower *Lactobacillus* count when stored at 10°C [59]. Furthermore, milk heated at 115° C for 2 s both with and without nisin was microbiologically acceptable after 28 days, even though milk with added nisin tasted better, with no off-flavors after 32 days. Similarly, no microbial growth was observed in milk treated concurrently with nisin (75–150 IU mL/L) and heat (117°C for 2 s) after a year of storage at 10 or 20°C.

3.8.5 Nisin in Combination with Nonthermal Treatments

Thermal treatments are now known to cause negative changes in the nutritional, sensory, and/or technological properties of milk. Because of nisin's antimicrobial effect against many spoilage and pathogenic microbes, innovative nonthermal food preservation resulted in microorganism inhibition with the minimal impact on quality and nutritional factors. Combining nisin with other antimicrobial agents or nonthermal treatments such as high pressure (HP) or high intensity pulsed electric field (HIPEF) has a synergistic effect in reducing microbial load, including bacterial spores [53].

3.8.6 Nisin with Other Antimicrobial Substances

Nisin, when combined with other antimicrobial compounds such as monolaurin, the lactoperoxidase system (LPS), or other bacteriocins can strengthen sensitization of resistant spoilage and foodborne microbes. Because of its antimicrobial properties, the monoester of lauric acid (Monolaurin) has gained significant attention. These properties may be enhanced when combined with nisin. An active action against bacilli could be presented in milk by combining monolaurin and nisin; specifically, the inhibition of *B. licheniformis* by both antimicrobial compounds increased with increasing pH when they were mixed together with milk. Furthermore, the combination of both compounds demonstrated bactericidal activity against different *Bacillus* species in skim milk, inhibiting regrowth and sporulation [38]. However, high concentrations of monolaurin may produce an unfavorable soapy odour and taste, which is undesirable in dairy products.

The LPS system in raw milk promotes storage stability at room temperature. The combination of LPS and nisin had a synergistic effect as well as long-lasting inhibitory activity on *L. monocytogenes* in reconstituted skim milk; additionally, its effectiveness was not affected by pH.

Both LPS and nisin have different modes of action, but they both cause cytoplasmic membrane damage, which explains their synergistic activity. The primary LPS reaction product, hypothiocyanite, is known to interact with the thiol groups of different proteins and inactivate critical enzyme and protein systems [4]. Nisin, on the other hand, forms a wedge-like pore from numerous nisin molecules bound to the membrane phospholipids. Both antimicrobials play a significant role in depleting proton motive force because LPS targets the bacterial respiration chain [4] and nisin pores result in a collapse of the membrane electrical potential and the gradient of pH.

Lysozyme, which is added to cheese milk to inhibit the activity of *Bacillus* spp., inhibited the activity of various lactobacilli strains when exposed to nisin, but did not affect nisin-producing lactococci used as a starter culture. Reuterin is a broad-spectrum antimicrobial substance formed by some *L. reuteri* strains. In milk, reuterin and nisin had a synergistic effect against *L. monocytogenes* and *S. aureus* [1].

Bacteria are undoubtedly present in the microbial flora of minimally processed dairy products, whereas molds and yeasts are commonly associated with special ingredients, such as wild blueberries added to yogurt. As a result, yogurt with minimally processed blueberries has a limited shelf life; thus, some phytopreservatives, such as vanillin, have been investigated for fungi removal in yogurt with blueberries, while added nisin cannot prevent spoilage [41]. Nisin significantly increased the viability of spoilage microbes in yogurt containing fresh peaches. This fact may create a disadvantage in the use of nisin because, in particular, yeast growth may be inhibited.

3.8.7 Nisin with High-Intensity Pulsed Electric Fields

High-intensity electric pulse (HIPEF) has received a lot of attention as a viable nonthermal technology for processing liquid foods. A few studies have been carried out to investigate the inactivation of milk microbes using this method of treatment. It was discovered that increasing the electric field intensity with the number of pulses and the nisin concentration have a synergetic step for inactivating *L. innocua* in skim milk.

Similar behavior was observed in the loss of viability with cell death of *S. aureus* in skim milk as a result of both treatments being applied concurrently, resulting in a remarkable synergistic action when compared to those treatments being applied individually [52].

Another hitch, such as a mild thermal treatment or other antimicrobials, may increase the synergy relationship between HIPEF treatment and nisin. The effect of a HIPEF-nisin treatment on the vegetative cells of *B. cereus* in milk may be enhanced by the addition of carvacrol, but only at high concentrations.

When *L. innocua* was observed using transmission electron microscopy, researchers discovered a remarkable morphological difference between nisin-HIPEF-treated cells and those treated with nisin and HIPEF alone as the electric field intensity continued to increase. Both nisin and HIPEF treatment resulted in cytoplasmic clumping and an increase in the roughness of the cell wall surface and cell membrane thickness, separation of the cytoplasm from the cell membrane, and finally a slight increase in the length of the cell. Therefore, combining nisin and HIPEF in a treatment method results in damage to the cell wall and cell membrane.

Gallo et al. [18] updated the sequence of applying nisin and HIPEF and came to the conclusion that applying nisin before HIPEF treatment forced their simultaneous effect, whereas adding nisin after HIPEF had no effect on the final action. This behavior suggests that any media variation, as well as those inflicted on the cell envelope by HIPEF, may impair nisin action. These research results partially agree with the mode of action of nisin and its amphipathic property, which allows nisin to interact with phospholipid membranes. Terebiznik et al. [54] claimed that no stronger effect of nisin could be expected when HIPEF has lethal actions.

3.8.8 Restrictions on Introducing Nisin in Dairy Products

There are several limitations to using nisin in dairy products, including its adsorption to fats and the surface of protein globules, heterogeneous distribution in dairy product matrices, inactivation of nonresistant starter cultures, and alteration of taste when nisin-producing strains are presented. To circumvent these constraints, nisin was encapsulated in phospholipid vesicles and tested in Cheddar cheese [34]. Although the fat content of milk is affected by the stability of liposome vesicles, vesicle-binding nisin has been shown to successfully withstand the Cheddar cheese-making temperature cycle and improve nisin stability, validity, and distribution [34]. The appearance of resistant cells in nisin-sensitive strains may constitute a further limitation of its use.

Because milk is a complex mixture of many different substances, including water, proteins, and fat, the validity of nisin in pasteurized milk may have been compromised by its composition. Several studies have found interactions between milk fat and nisin activity, which may limit its use in fat-containing dairy products. According to Jung et al. [27], increasing the concentrations of milk fats reduced nisin activity against *L. monocytogenes*. Emulsifiers may be useful in improving nisin activity; polyoxyethylene sorbitan monooleate, also known as Tween 80, was tested in half/whole milk and found to retain 43.4% of the original nisin activity after 2 h at 37°C, while no emulsifier retained only 19.6%.

3.8.9 Nisin and COVID-19

With over 18 million cases of infection worldwide, the ongoing global outbreak of COVID-19, a severe life-threatening infectious respiratory disease caused by a recently discovered severe acute respiratory syndrome coronavirus 2 (SARS-CoV-2), has severely impacted human life (https://coronavirus.jhu.edu/map.html). There is currently no specific antiviral medication available for COVID-19, but extensive efforts are being made around the world. Although vaccines are thought to be the most powerful weapon in the fight against virus invasion, they may take a long time to make it from the lab to successful human applications. Given the acute nature of the COVID-19 pandemic, there is an urgent need to develop effective antiviral therapeutics for COVID-19 prevention and treatment. It is widely acknowledged that the spike protein on the outer surface of SARS-CoV-2 is a critical recognition factor for its attachment and entry into host cells. The binding of the RBD (receptor binding domain) of spike protein to the human angiotensin-converting enzyme 2 (hACE2) receptor initiates viral infection in humans. As a result, a therapeutic agent that inhibits hACE2 could prevent the interaction of SARS-CoV-2 spike protein and thus reduce infection establishment. Although small non-proteinaceous molecules are commonly used as therapeutics, they are ineffective at blocking protein-protein interactions (PPIs), especially when a deep binding pocket at the interface is missing. Peptides, on the other hand, are better suited for disrupting PPIs by specifically interacting with the interfaces. More importantly, small peptides have decreased immunogenicity. As a result, peptides may be ideal candidates for use as novel therapeutics. The recently described peptides are all small, synthetic, and expensive, and they have not shown promising results against SARS-CoV-2. The peptides recently designed computationally against the SARS-CoV-2 must be synthesized before they can be used in practice, so they are not natural or food grade. Among all nisin variants tested, nisin Z, nisin A, nisin U, and nisin H were the most effective at interacting with human endothelial cell surface-receptor hACE2, the site where RBD of SARS-CoV-2 spike binds to initiate infection. Nisin has a higher affinity for the hACE2 receptor than the RBD of the viral spike protein. Because nisin is a low molecular weight peptide that is readily bioavailable in the system, its binding to hACE2 is expected to override the interaction possibility of the RBD of spike of SARS-CoV-2 and effectively prevent the virus from entering the host cell. Because nisin is a heat-stable natural food grade peptide that can be produced cost-effectively, even in large quantities, through microbial fermentation, the current study will pique researchers' interest in developing a new nisin-based treatment strategy for COVID-19, either through oral or nasal applications. However, additional experimental validation is required to determine its doses and mechanistic application to test the competition of nisin and spike protein of SARS-CoV-2 for human access [42].

References

1. Arques, J. L., Fernandez, J., Gaya, P., Nunez, M., Rodrıguez, E., and Medina, M. (2004). Antimicrobial activity of reuterin in combination with nisin against food-borne pathogens. Int. J. Food Microbiol., 95: 225–229.
2. Balciunas, E. M., Martinez, F. A. C., Todorov, S. D., de Melo Franco, B. D. G., Converti, A., and de Souza Oliveira, R. P. (2013). Novel biotechnological applications of bacteriocins: A review. Food Control, 32(1): 134–142.
3. Benech, R.-O., Kheadr, E. E., Lacroix, C., and Fliss, I. (2002). Antibacterial activities of nisin Z encapsulated in liposomes or produced in situ by mixed culture during Cheddar cheese ripening. Appl. Environ. Microbiol., 68: 5607–5619.
4. Boots, J.-W. and Floris, R. (2006). Lactoperoxidase: From catalytic mechanism to practical applications. Int. Dairy J., 16: 1272–1276.
5. Chang, Y. (2020). Bacteriophage-derived endolysins applied as potent biocontrol agents to enhance food safety. Microorganisms, 8: 724. doi: 10.3390/microorganisms8050724.

6 Cintas, L. M., Casaus, M. P., Herranz, C., Nes, I. F., and Hernández, P. E. (2001) Review: Bacteriocins of lactic acid bacteria. Int. Food Sci. Technol., 7(4): 281–305.

7. Cleveland, J., Montville, T. J., Nes, I. F., and Chikindas, M. L. (2001). Bacteriocins: Safe, natural antimicrobials for food preservation. Int. J. Food Microbiol., 71(1): 1–20. https:// doi.org/10.1016/S0168-1605(01)00560-8.

8. Cotter, P. D., Ross, R. P., and Hill, C. (2012). Bacteriocins — A viable alternative to antibiotics? Nat. Rev. Microbiol., 11: 95–105. https://doi.org/10.1038/nrmicro2937.

9. Cotter, P. D., Hill, C., and Ross, R. P. (2005). Bacteriocins: Developing innate immunity for food. Nat. Rev. Microbiol., 3: 777–788.

10. Daba, G. M. and Elkhateeb, W. A. (2020). Bacteriocins of lactic acid bacteria as biotechnological tools in food and pharmaceuticals: Current applications and future prospects. Biocatal. Agri. Biotechnol., 28: 101–750.

11. Davies, E. A., Bevis, H. E., Potter, R., Harris, J., Williams, G. C., and Delves-Broughton, J. (1998). Research note: The effect of pH on the stability of nisin solution during autoclaving. Lett. Appl. Microbiol., 27(3): 186–187. https://doi.org/10.1046/ j.1472-765X.1998.t01-1-00401.x.

12. de Almeida, G. M., Carlos, M. N. M., Attilio, C., and Ricardo, P. S. O. (2020). Kinetic and thermodynamic parameters of nisin thermo-inactivation. J. Food Eng. 280: 109986. https://doi.org/10.1016/ j.j.foodeng.2020.109986.

13. Delves-Broughton, J. (1990). Nisin and its application as a food preservative. Int. J. Dairy Technol., 43(3): 73–76. https://doi.org/10.1111/j.1471-0307.1990.tb02449.x.

14. De Vuyst, L. and Leroy, F. (2007). Bacteriocins from lactic acid bacteria: Production, purification, and food applications. J. Mol. Microbiol. Biotechnol., 13: 194–199.

15. Devi, M. P., Bhowmick, N., Bhanusree, M. R., and Ghosh, S. K. 2015. Preparation of value-added products through reservation. Springer India, 1: 113–124.

16. Dimov, S.; Ivanova, P., and Harizanova, N. (2005) Genetics of bacteriocin biosynthesis by lactic acid bacteria. J. Biotechnol., 19(3): 3–10.

17. Drider, D., Fimland, G., Hechard, Y., Mcmullen, L.M., and Prevost, H. (2006). The continuing story of class IIa bacteriocins. Microbiol. Mol. Biol. Rev., 70(2): 564–582.

18. Gallo, L. I., Pilosof, Am. M. R., and Jagus, R. J. (2007). Effect of the sequence of nisin and pulsed electric fields treatments and mechanisms involove in the inactivation of Listeria innocua in whey. Journal of Food Engineering, 79, 188–193.

19. Gálvez, A., Abriouel, H., Lόpez, R. L., and Ben, O. N. 2007. Bacteriocin-base strategies for food biopreservation. Int. J. Food Microbiol., 120: 51–70.

20. Galvez, A., Lopez, R. L., Abriouel, H., Valdivia, E., and Omar N. B. (2008). Application of bacteriocins in the control of foodborne pathogenic and spoilage bacteria. Crit. Rev. Biotechnol., 28(2): 125-152.

21. Ganguly, S. (2013). Basic principles for effective food preservation: A review. Int. J. Pure Appl. Biosci., 1: 84–85.

22. Gratia, A. (1925). Sur un remarquable exemple d'antagonisme entre deux souches de coilbacille. CR Seances Soc. Biol. Fil, 93: 1040–1041.

23. Gulluce, M., Karadayi, M., and Baris, O. (2013) Bacteriocins: Promising natural antimicrobials. Sci. Technol. Edu., 4(2): 1016–1027.

24. Hammami, R., Fernandez, B., Lacroix, C., and Fliss, I. (2013) Anti-infective properties of bacteriocins: An update review. Cell. Mol. Life Sci., 70(16): 2947–2967.

25. Hsu, S. T. D., Breukink, E., Tischenko, E., Lutters, M. A. G., de Kruijff, B., Kaptein, R., Bonvin, A. M. J. J., and van-Nuland, N. A. J. (2004). The nisin lipid II complex reveals a pyrophosphate cage that provides a blueprint for novel antibiotics. Nat. Struct. Mol. Biol., 11(10): 963–967.

26. Jacob, F., Lwoff, A., Siminovitch, A., and Wollman, E., (1953). Definition of some terms relative to lysogeny. Ann. Inst. Pasteur., 84(1): 222–224.

27. Jung, D. S., Bodyfelt, F. W., and Daeschel, M. A. (1992). Influence of fat and emulsifiers on the efficacy of nisin in inhibiting Listeria monocytogenes in fluid milk. J. Dairy Sci., 75: 387–393.

28. Kannan, K. and Mankin, A. S. (2011). Macrolide antibiotics in the ribosome exit tunnel: Species-specific binding and action. Ann. N Y Acad. Sci., 1241: 33–47.

29. Kim, K. P., Klumpp, J., and Loessner, M. J. 2007. Enterobacter sakazakii bacteriophages can prevent bacterial growth in reconstituted infant formula. Int. J. Food Microbiol., 115: 195–203.

30. Kjos, M., Nes, I. F., and Diep, D. B. (2009). Class II one-peptide bacteriocins target a phylogenetically defined subgroup of mannose phosphotransferase systems on sensitive cells. J. Microbiol., 155: 2949–2961.

31. Klaenhammer, T. R. (1993). Genetics of bacteriocins produced by lactic acid bacteria. FEMS Microbiol. Rev., 12(1–3): 39–85.

32. Kleanthous, C. (2010). Swimming against the tide: Progress and challenges in our understanding of colicin translocation. Nat. Rev. Micro., 8(12): 843–848.

33. Kleerebezem, M. and Quadri, L. E. (2001) Peptide pheromone-dependent regulation of antimicrobial peptide production in Gram-positive bacteria: A case of multicellular behavior. J. Peptides, 22(10): 1579–1596.

34. Laridi, R., Kheadr, E. E., Benech, R. O., Vuillemard, J. C., Lacroix, C., and Fliss, I. (2003). Liposome encapsulated nisin Z: Optimization, stability and release during milk fermentation. Int. Dairy J., 13: 325–336.

35. Lin T.-H. and Pan T.-M. (2017). Characterization of an antimicrobial substance produced by Lactobacillus plantarum NTU 102. J. Microbiol. Immunol. Infect., 52: 1–9.

36. Lins, L., Ducarme, P., Breukink, E., and Brasseur, R. (1999). Computational study of nisin interaction with model membrane. Biochim. Biophys. Acta, 1420: 111–120.

37. Macwana, S. and Muriana, P. M. (2012). Spontaneous bacteriocin resistance in Listeria monocytogenes as a susceptibility screen for identifying different mechanisms of resistance and modes of action by bacteriocins of lactic acid bacteria. J. Microbiol. Methods, 88(1): 7–13.

38. Mansour, M. and Milliere, J. B. (2001). An inhibitory synergistic effect of a nisin–monolaurin combination on Bacillus sp. vegetative cells in milk. Food Microbiol., 18: 87–94.

39. Mauriello, G., Ercolini, D., La Storia, A., Casaburi, A., and Villani, F. (2004). Development of polythene films for food packaging activated with an antilisterial bacteriocin from Lactobacillus curvatus 32Y. J. Appl. Microbiol., 97: 314–322. doi:10.1111/j.1472-765X.2005.01796.x.

40. Mitra, S., Mukhopadhyay, B. C., and Biswas, S. R. (2011). Potential application of the nisin Z preparation of Lactococcus lactis W8 in preservation of milk. Lett. Appl. Microbiol., 98–105. https://doi.org/10.1111/j.1472-765X.2011.03075.x.

41. Penney, V., Henderson, G., Blum, C., and Johnson-Green, P. (2004). The potential of phytopreservatives and nisin to control microbial spoilage of minimally processed fruit yoghurts. Innov. Food Sci. Emerg. Technol., 5: 369–375.

42. Bhattacharya, R., Gupta, A. M., Mitra, S., Mandal, S., and Biswas, S. R. (2021). A natural food preservative peptide nisin can interact with the

SARS-CoV-2 spike protein receptor human ACE2. Virology, 552: 107–111. https://doi.org/10.1016/j.virol.2020.10.002.

43. Rao, K. V. S. S. and Mathur, B. N. (1996). Thermal death kinetics of Bacillus stearothermophilus spores in a nisin supplemented acidified concentrated buffalo milk system. Milchwissenschaft, 51: 186–191.

44. Riley, M. A. and Chavan, M. A. (2007). Bacteriocins: Ecology and Evolution. Springer-Verlag Berlin Heidelberg, p. 150.

45. Riley, M. A. and Wertz, J. E. (2002). Bacteriocins, evolution, ecology and application. Annu. Rev. Microbiol., 56: 117–137. https://doi.org/10.1146/annurevmicro.56.012302.161024.

46. Rodrıguez, E., Arques, J. L., Gaya, P., Tomillo, J., Nunez, M., and Medina, M. (2000). Behaviour of Staphylococcus aureus in semi-hard cheese made from raw milk with nisin-producing starter cultures. Milchwissenschaft, 55: 633–635.

47. Rogers, L. A. (1928). The inhibitory effect of Streptococcus lactis on Lactobacillus bulgaricus. J. Bacteriol., 16(2): 321–325.

48. Scannell, A. G., Hill, C., Ross, R. P., Marx, S., Hartmeier, W., and Arendt, E. K. (2000). Development of bioactive food packaging materials using immobilised bacteriocins lacticin 3147 and nisaplin. Int. J. Food Microbiol., 60: 241–249.

49. Sebti, I., Delves-Broughton, J., and Coma, V. (2003). Physicochemical properties and bioactivity of nisin-containing cross-linked hydroxypropylmethylcellulose films. J. Agric. Food Chem., 51: 6468–6474.

50. Simons, A., Kamel, A., and Duval, R. E. (2020). Bacteriocins, antimicrobial peptides from bacterial origin: Overview of their biology and their impact against multidrug-resistant bacteria. Microorganisms, 8(5). doi:10.3390/microorganisms8050639.

51. Singh, V. P. 2018. Recent approaches in food bio-preservation - A review. Open Vet. J., 8: 104–111.

52. Sobrino-Lopez, A., Raybaudi-Massilia, R., and Martın-Belloso, O. (2006). Enhancing inactivation of Staphylococcus aureus in skim milk by combining high intensity pulsed electric fields and nisin. J. Food Prot., 69: 345–353.

53. Sobrino-López, A. and Martín-Belloso, O. (2008). Use of nisin and other bacteriocins for preservation of dairy products. Int. Dairy J., 18(4): 329–343.

54. Terebiznik, M., Jagus, R., Cerrutti, P., de Huergo, M. S., and Pilosof, A. M. R. (2000). Combined effect of nisin and pulsed electric fields on the inactivation of Escherichia coli. J. Food Prot., 63: 741–746.

55. Thomas, L. V., Delves-Broughton, J. (2005). Nisin. In: Davidson, P. M., Sofos, J., Branen, A. L. (Eds.), Antimicrobials in Food, 3rd ed. CRC Press, Boca Raton, FL, pp. 237–274.

56. Thomas, L. V., Clarkson, M. R., and Delves-Broughton, J. (2000). Nisin. In: Naidu, A. S. (Ed.), Natural Food Antimicrobial Systems. CRC Press, Boca Raton, FL, pp. 463–524.

57. Ulpathakumbura, C. P., Senaka, R. C., Senavirathne, N. D., Jayawardene, L. P. I. N. P., Prasanna, P. H. P., Janak, K., Vidanarachchi, C. (2016). Effect of biopreservatives on microbial, physico-chemical and sensory properties of Cheddar cheese. Food Biosci., 13: 21–25. http://dx.doi.org/10.1016/j.fbio.2015.12.003.

58. Vriezen, J. A. C., Valliere, M., and Riley, M. A. (2009). The evolution of reduced microbial killing. Genome Biol. Evol., 1: 400–408.

59. Wirjantoro, T. I. and Lewis, M. J. (1996). Effect of nisin and high temperature pasteurization on shelf life of whole milk. J. Soc. Dairy Technol., 49: 99–102.

Chapter 4

New Unconventional Preservatives

Ghada M. Youssef[a] and Youssef M. Riyad[b]
[a]*Agricultural Research Center, Food Technology Research Institute, Special Food and Nutrition Department, Giza, Egypt*
[b]*Cairo University Faculty of Agriculture, Food Science Department, Giza, Egypt*
y-mriyad@hotmail.com

Frankincense (*Boswellia carterii*) water extracts and squid ink are used as new unconventional preservatives. They contain different active compounds. The essential oil of lavender leaves (Lavandula latifolia) and lemon balm leaves (Melissa officinalis L) are aromatic oily liquids. Essential oils have been widely used as food flavors and preservatives. In recent years, essential oils have been qualified as natural antioxidants and proposed as potential substitutes for synthetic antioxidants in food preservation.

4.1 Introduction

Preservatives are any chemicals that lead to delay or prevent deterioration when added to the food. Frankincense (*B. carterii*)

Natural Food Preservatives
Edited by Sourish Bhattacharya and Heba Hassan Abd-El Azim Salama

ISBN 978-981-4968-64-5 (Hardcover), 978-1-003-36776-5 (eBook)
www.jennystanford.com

water extracts and squid ink are used as new unconventional preservatives. They contain different active compounds.

4.2 Frankincense

Frankincense is a resin obtained from the trees belonging to the genus *Boswellia* (family Burseraceae). It has orange-brown color and a direct relationship was found between levels of pigments and phenolic compounds [1, 2]. There are many species and varieties of frankincense trees, such as *B. carterii*, *B. serrata*, *B. frereana*, and *B. sacra* in East Africa, China, India, Somalia, and Arabia, respectively. The differences between frankincense resin species depend on soil and climate. The frankincense tree plays an important role in the economy of many Middle Eastern countries. It is also grown in many African countries such as Somalia, Sudan, and southern Egypt [3].

Frankincense has many medical benefits as it is used in folk medicine and in the treatment of many diseases such as tumors, ulcers, leprosy, dysentery, and chest diseases such as coughing, asthma, and distressed breathing; it is also used as a tonic for the heart [4, 5].

Boswellic resin improves memory and learning in the treated Alzheimer's group, modulates the oxidative stress, is involved in the protective effect against amyloid deposition and neurodegeneration, and stimulates the immune system in the mouse brain [6].

Frankincense is used in most countries of the world. Frankincense is one of the most famous materials that are sold and found in aromatherapy shops and Arab medicinal herbs. Frankincense is used to treat many diseases. In addition, it has very slight side effects and is a cheap natural source, especially when compared to the side effects and high prices of some antibiotics [7].

Frankincense is commonly used in pharmaceutical products because it contains active compounds and is useful in treating many human diseases. The taste of frankincense is bitter, so few people like to chew frankincense. There is little research on frankincense in supporting foods. Among the nutritional applications is the use of frankincense in supporting broiler chicken diets with *B. serrata* resin rich in biologically active ingredients as this application led to improved productivity, chemical composition, and nutritional value of the meat produced [8].

Frankincense (*B. carterii*) has many medicinal benefits. In Egypt, some people like to use it in chewing and sometimes in incense. Frankincense powder was prepared by soaking in drinkable water for 24 h (T1) and boiling in water for 1 h (T2) to the fortification of guava nectar in concentrations ranging from 0.25% to 2% (volume/volume (v/v)). The phenols and total flavonoid contents of the T2 sample were 0.869 and 0.147 mg/g, respectively. However, it was 0.748 and 0.070 mg/g, respectively, for T1. Frankincense powder has 93.74% antioxidant activity with 67.26 mg/mL IC_{50}. The T2 sample has antibacterial activity against *Salmonella typhi* and *Bacillus cereus*. Frankincense powder has cytotoxic activity and this activity is strong on HepG-2 human cancer cell lines compared to the positive control (doxorubicin); however, it is less activity on the human breast adenocarcinoma (MCF-7). The IC_{50} values of the frankincense powder were 7.8 and 33.9 μM, respectively, while doxorubicin values were 10.3 and 28.5 μM, respectively. It is possible to mix aqueous extracts of the frankincense with guava nectar to obtain a product with a pleasant taste, color, and aroma and has also the advantages of frankincense [9].

4.3 Squid Ink

The squid belongs to the order Teuthida in cephalopods. It has eight arms and longer tentacles that help it to swim. It tends to stay in cold and deep water. It is black and can be found in several sizes. These species are not more than 60 cm in length. In some countries, it is known as the calamari [10].

Cephalopod species create ink which helps it escape from predators. The ink is diffused like a cloud as a screen to help hide him a little. Therefore, it is necessary for the ink to have dark properties and thick stability to prevent it from dispersing quickly into the surrounding water. The color of ink differs in each species of this family and is known by many names such as squid ink, cuttlefish ink, tinta calamar, nero di seppia, black squid ink, cephalopod ink, and octopus ink. Octopus ink is usually black ink, while squid color is blue-black and squid ink is like brown shade. In some countries, they prefer squid ink because of its unique appearance [11].

Squid ink is very high in several important nutrients, particularly the antioxidants, which are well-known to help in protecting the cells and the heart against damage from free radicals. The noticeable blue-black color of squid ink is due to the presence of large amounts of melanin pigment. The melanin pigment is produced in mature cells of the ink gland which is present in the bottom of the ink sac. Additionally, it also contains large quantities of proteins, lipids, minerals, and so on [12]. Squid ink has better antimicrobial activity against various microbes, making it a great choice of natural antibiotic [13].

Cuttlefish ink extract and freshwater mussel extract had no toxic effect on experimental rats up to a concentration of 2000 mg/kg. The LD_{50} in mice was therefore taken as above 2000 mg/kg. This dose was considered one-tenth of the proposed LD_{50}, that is, 200 mg/kg body weight [14].

Cuttlefish ink is usually discarded as a processing waste but is common in the cuisine of Japan, Italy, and Spain. Surprisingly, there are few available studies regarding its proximate composition which is necessary to ensure that it meets the nutritional requirements essential for the growth and health of a living being. Food with high protein is generally considered to be nutritionally better, and hence ink from *Sepia prabahari* with the highest protein content is nutritionally preferred among the three cuttlefish species *S. pharaonis*, *S. prabahari*, and *S. ramani* [15].

Loligo duvauceli ink generated as a by-product in the processing industry has a large number of compounds such as phenol, flavonoid, protein, carbohydrate, and amino acids. The protein and flavonoids were present in the higher amount, whereas sterols, terpenoids, and saponins were absent [16].

Squid ink, which is discarded as a by-product, could be used as a good source of natural antioxidants after melanin removal. Melanin-free ink with high thermal stability possessed radical scavenging and metal chelating activities. After ultrafiltration, the fraction with lower molecular weight had a greater antioxidative activity [17].

Squid ink has many uses in the food industry owing to its distinct flavor. Squid ink is sold in commercial markets to be used as a food flavoring [18]. In many dishes, squid ink is used as a food coloring, for example, black rice, baby squid in ink sauce, ink soup with pork and squid and imitation caviar [19]. Squid ink is an ingredient in bread, sweets, tofu, pasta (spaghetti), curry, potato chips, sweets, snacks [20], and sauces [21].

Squid ink is disposed of during squid preparation, which may cause environmental pollution and health problems. Squid ink had antimicrobial activity against *B. subtilis, Aspergillus fumigatus, A. niger, Candida albicans, Staphylococcus aureus, Escherichia coli, Klebsiella pneumonia*, and *Proteus vulgaris* with zone inhibitions of 19, 18, 12, 13, 11, 13, 12, and 10 mm, respectively. Squid ink has an antioxidant activity which was 91.66%. The results of the cell viability and toxicity assay of squid ink indicated that as the concentration of squid ink increased, the viability percentages decreased, while inhibitory activity percentages increased. The IC_{50} [the concentration causing the death of 50% of the human tumor cell line (lung carcinoma cell line A549)] was 22 μg/mL [22].

4.4 Lavender (*L. latifolia*) Leaves Essential Oil

Essential oils, also called volatile odoriferous oil, are aromatic oily liquids extracted from different parts of plants, for example, leaves, peels, barks, flowers, buds, and seeds. They can be extracted from plant materials by steam distillation, water distillation, expression, and other methods. Essential oils have been widely used as food flavors [23, 24]. Essential oils from plants have been known to act as natural additives and as antimicrobial and antioxidant agents. Their activities vary with the source of plants, chemical composition, and extraction methods. In recent years, essential oils have been qualified as natural antioxidants and proposed as potential substitutes for synthetic antioxidants in food preservation. The safety of synthetic antioxidants has been doubted [25].

The lavender is a genus of about 25–30 species of plants in the Lamiaceae family, and the Mediterranean region is the native. The genus includes annuals, subshrubs, small shrubs, and herbaceous plants [26]. Lavender essential oil is popular as complementary medicine, anti-depressive and anti-inflammatory properties, in addition to its recognized antimicrobial effects [27]. The pleasant aroma of lavender is mainly due to the occurrence of low molecular weight terpenoids synthesized and accumulated in aerial parts, especially in inflorescences. The distilled volatile oil of this plant has gained great importance in aromatherapy and in perfume, cosmetic and flavoring industries since antiquity. Furthermore, volatile constituents of lavender are of special significance in the pharmaceutical and food industries [28].

The main active ingredients of lavender leaves essential oil are monoterpenes (linalool, linalyl acetate, lavandulol, and geraniol); these components have different antibacterial and antifungal effects and antioxidant activity, depending on their chemical composition [29]. A high and almost equal content of linalool and linalyl acetate is required for good antimicrobial properties of lavender essential oil. The composition of any essential oil depends on several factors, such as local, climatic, seasonal, different stages of plant maturity, and experimental conditions [30].

The acute oral toxicity using rats was over 2000 mg/kg of body weight (bw) for lavender (*L. angustifolia*) essential oils and the acute dermal value was over 4000 mg/kg bw. Also, lavender has no irritation for skin and eyes [31].

There are many plants that are essential oil sources, around 300 are of commercial value. Approximately 30 of these plants are cultivated on a large scale, produce essential oils whose use as drugs dates back centuries, and are described in international and national Pharmacopoeias. Lavender essential oil is listed in European and Pharmacopoeias [32].

The bioactive components of lavender leaves were studied and used to prolong the shelf life of the cake. Total phenols content in lavender leaves powder was higher than that of lavender leaves water extract (15.74 vs. 2.36 mg gallic acid equivalent/g). The lavender leaves essential oil exhibited the highest antioxidant activity (85.32%), followed by lavender leaves powder (76.71%) and lavender leaves water extract (27.38%). Also, lavender leaves essential oil has activity against pathogenic bacteria such as *S. aureus*, *B. cereus*, *E. coli*, and *S. typhi*, and it has a strong antitumor activity against intestinal carcinoma cell line (Caco-2). The results of the sensory evaluation showed that the best characteristics were obtained by using 600 ppm of essential oil, 10% of the water extract, or 7.5% of the leaves powder. In conclusion, it is possible to extend the shelf life of the cake to 28 days by addition of 10% of the lavender leaves water extract or 7.5% of lavender leaves powder to cake ingredients or to 4 months by addition of 600 ppm of lavender leaves essential oil compared to 7 days for control sample at room temperature and also with good sensory properties for the produced cake [33].

4.5 *M. officinalis* L. Leaves Oil

M. officinalis L. oil is a perennial edible herb native to the Mediterranean region. The plant is cultivated in various parts of the world and grows especially in western Asia, southwestern Serbia, and North Africa. It is considered an important medicinal plant and is largely used in traditional medicine for the treatment of headaches, indigestion, colic, nervousness, cardiac failure, and depression [34].

The essential oil has high hydrophobicity, but can diffuse, dissolve and permeate into the most common polymers [35]. Therefore, the application of essential oils can have great potential in polymer systems.

M. officinalis L. oil possesses antibacterial properties and has the potent antibacterial properties against some representative foodborne pathogens. Therefore, they could be used as possible food antimicrobial preservatives in the food industry, but in vivo studies should be done to evaluate the probable adverse effects on food sensory properties [36].

M. officinalis L. oil showed antibacterial effect against *S. aureus, B. cereus, B. subtilis, B. pumilus, Pseudomonas aeruginosa, S. poona, E. coli*, and *L. innocua*. However, lemon balm oil activity against fungi Fusariumoxy sporumalbedinis, F. oxyxporumlini, and Mucorramannianus was also found [37].

M. officinalis L. is a safe plant without any pathological symptoms if used in concentrations of 500–600 mg and has many uses in the industry such as in medicine, perfume, cosmetics, and food [38, 39].

Tofu is the most well-known and common soya food, and it has been consumed for may be 2000 years in Asia. It has become well-known and respected in much of the world during the last 20 years. It is similar to the existing dairy milk cheese products, such as paneer and peynir in India and Turkey, and is processed in the same way; it provides an excellent soy protein alternative to many traditional foods. The edible coating was used to extend the shelf life of food products and act as selective barriers to oxygen uptake, moisture transfer, lipid oxidation, and losses of volatile aromas and flavors [40].

The application of edible coating incorporated with essential oil on tofu and strawberry was found to be effective in delaying the microbial degradation of tofu and strawberry with minimum inhibition concentration which will be incorporated with an edible coating, the agent was 0.6% lemon peel essential oil for tofu and 1% lemon peel essential oil for strawberries [41]. Incorporating antimicrobial compounds into edible films or coatings provides a novel way to improve the quality and shelf life of ready-to-eat foods [42].

We still need new methods of reducing or eliminating food spoilage and foodborne pathogens, possibly in combination with the existing approaches [43]. The demand for nontoxic, natural preservatives has been rising with increased awareness and reports of the ill effects of synthetic chemicals present in food. Furthermore, the emergence of foodborne pathogens has lately become a major public health concern [44].

Fresh tofu samples were immersed in different solutions containing gelatin (1% and 2%) and *M. officinalis* L. oil (0.1%, 0.3%, and 0.5%) as antimicrobial activity. The samples were stored up to 14 days under cooling conditions (4°C). *M. officinalis* L. oil was chemically and microbiologically analyzed. Rheological properties and thixotropic effect were determined. The blends of gelatin and *M. officinalis* L. oil exhibited dilatant flow behavior and all samples showed a thixotropic effect. The effect of edible coating on weight loss, moisture content, texture properties, microbiological tests, and sensory evaluation of tofu samples was determined during the storage period as an indication of shelf life and quality of tofu samples [45].

References

1. Pietta, P. G. Flavonoids as antioxidants. Journal of Natural Products 2000, 63(7), 1035–1042.
2. Elham, G.; Reza, H.; Jabbar, K.; Parisa, S.; Rashid, J. Isolation and structure characterisation of anthocyanin pigments in black carrot (*Daucus carota* L). Pakistan Journal of Biological Sciences 2006, 9(15), 2905–2908.
3. Maloney, G. A. Gold Frankincense & Myrrh: An Introduction to Eastern Christian Spirituality, 1st edition. Crossroad Publishing Company, New York, 1997; p. 178.
4. Bonjar, S. Evaluation of antibacterial properties of some medicinal plants used in Iran. Journal of Ethnopharmacology 2014, 94, 301–305.

5. Aldory, M. E.; Ali, F. F.; Sultan, S. M. Effective of watery and alcoholic extract of frankincense on the Candida albicans fungus. International Journal of Pharmaceutical Research and Allied Sciences 2018, 7(3), 56–62.

6. Zerrouki, K.; Djebli, N.; Gadouche, L.; Orhan, E. I.; Deniz, S. F.; Erdem, A. S. Protective effect of boswellic resin against memory loss and Alzheimer's induced by aluminum tetrachloride and D-galactose (experimental study in mice). Phytothérapie 2020, Online First, https://doi.org/10.3166/phyto-2020-0222.

7. Hamidpour, R.; Hamidpour, S.; Hamidpour, M.; Shahlari, M.; Hamidpour, R. Chemistry, Pharmacology and medicinal property of frankincense (*Boswellia* species): From the selection of traditional applications to the novel phytotherapy for the prevention and treatment of serious diseases. Global Journal of Medical Research B 2015, 15(3), 15–22.

8. Al-Yasiry, A. R. M.; Kiczorowska, B.; Samolińska, W. Effect of *Boswellia serrata* resin supplementation on basic chemical and mineral element composition in the muscles and liver of broiler chickens. Biological Trace Element Research 2017, 179, 294–303. doi: 10.1007/s12011-017-0966-6.

9. Riyad, Y. M.; Barakat, H. A.; Amer, S. A. Evaluation of the properties of frankincense powder and its water extracts and the effect of its addition on guava nectar characteristics. Egyptian Journal of Food Science 2020, 48 (1), 159–171.

10. Scientific American studies were conducted about "Can a Squid Fly Out of the Water". https://www.scientificamerican.com/article/can-squid-fly, last accessed on 15.01.2018.

11. Caldwell, R. L. An observation of inking behavior protecting adult Octopus bocki from predation by green turtle (Chelonia mydas) hatchlings. Pacific Science 2005, 59, 69–72.

12. Palumbo, A.; Di Cosmo, A.; Gedualdo, I.; Hearing, V. J. Subcellular localization and function of melanogenic enzymes in the ink gland of *Sepia officianalis*. Biochemical Journal 1997, 323, 749–756.

13. Nithya, M.; Ambikapathy, V.; Panneerselvam A. Effect of pharaoh's cuttlefish ink against bacterial pathogens. Asian Journal of Plant Science and Research 2011, 1(4), 49–55.

14. Fahmy, S. R.; Soliman, A. M. In vitro antioxidant, analgesic and cytotoxic activities of *Sepia officinalis* ink and *Coelatura aegyptiaca* extracts. African Journal of Pharmacy and Pharmacology 2013, 7(22), 1512–1522.

15. Ganesan, P.; Brita Nicy, A.; Kanaga, V.; Velayutham, P. Proximate analysis of cuttlefish ink procured from Thoothukudi coast: A comparative study. IJFAS 2017, 5(3), 253–255.
16. Nisha, N.; Suja, S. Phyto chemical evalution and antioxidant activity of methanol extract of *Loligo duvauceli* Ink. Journal of Pharmacognosy and Phytochemistry 2018, 7(1), 1764–1767.
17. Vate, N. K.; Benjakul, S. Antioxidative activity of melanin-free ink from splendid squid (*Loligo formosana*). International Aquatic Research 2013, 5, 9–20.
18. Hazan, M. Essentials of Classic Italian Cooking. Knopf Inc., New York, USA, 1992.
19. Marquinet, A.; Inaki, J. Process for producing a food colorant, colorant thus obtained and uses thereof. U.S. Patent 6, 329,010 B1, December 11, 2001.
20. Kim, H. S. Kimchi using sepia and soy sauce and method for manufacturing the same. Korean Patent 2009, 10, 2009-0079541.
21. Kim, M. J. India ink soy sauce additive of uncurled bean curd or bean curd stew, and its manufacturing method. Korean Patent 2008, 10, 2008-0108782.
22. Riyad, Y. M.; Rizk, A. E.; Mohammed, N. S. Active components of squid ink and food applications. Egyptian Journal of Food Science 2020, 48(1), 123–133.
23. Burt, S. Essential oils: Their antibacterial properties and potential applications in foods: A review. International Journal of Food Microbiology 2004, 94, 223–253.
24. Di Leo, L. P.; Retta, D.; Tkacik, E.; Ringuelet, J.; Coussio, J. D.; van Baren, C.; Bandoni, A. L. Essential oil and by-products of distillation of bay leaves (*Laurus nobilis* L.) from Argentina. Industrial Crops and Products 2009, 30, 259–264.
25. Darughe, F.; Barzegar, M.; Sahari, M. A. Antioxidant and antifungal activity of coriander (*Coriandrum sativum* L.) essential oil in cake. International Food Research Journal 2012, 19(3), 1253–1260.
26. Piccaglia, R.; Marotti, M.; Giovanelli, E.; Deans, S. G.; Eaglesham, E. Antibacterial and antioxidant properties of Mediterranean aromatic plants. Industrial Crops and Products 1993, 2, 47–50.
27. Da Silva, G.; Luft, C.; Lunardelli, A.; Amaral, R.H.; Melo, D. A. D.; Donadio, M. V. F.; Nunes, F. B.; De Azambuja, M. S.; Santana, J. C.; Moraes, C. M. B.; et al. Antioxidant, analgesic and anti-inflammatory effects of lavender

essential oil. Annals of the Brazilian Academy of Sciences 2015, 87(2), 1397–1408.

28. Cong, Y.; Abulizi, P.; Zhi, L.; Wang, X.; Mirensha, Y. Chemical composition of the essential oil of *Lavandula angustifolia* from Xinjiang, China. Chemistry of Natural Compounds 2008, 44(6), 108.

29. Janeczko, Z.; Pisulewska, E. Domestic Oil Plants. Occurrence, Cultivation, Chemical Composition, Application. "Know-How" Publishing, Kraków, Poland, 2008; pp. 7–11, 43–47.

30. Białoń, M.; Krzyśko-Łupicka, T.; Nowakowska-Bogdan, E.; Wieczorek, P. P. Chemical composition of two different lavender essential oils and their effect on facial skin microbiota. Molecules 2019, 24(18), 3270–3275.

31. Mekonnen, A.; Tesfaye, S.; Christos, S. G.; Dires, K.; Zenebe, T.; Zegeye, N.; Shiferaw, Y.; Lulekal, E. Evaluation of skin irritation and acute and subacute oral toxicity of *Lavandula angustifolia* essential oils in rabbit and mice. Journal of Toxicology 2019. https://doi.org/10.1155/2019/5979546

32. Sharifi-Rad, J.; Sureda, A.; Tenore, G. C.; Daglia, M.; Sharifi-Rad, M.; Valussi, M.; Tundis, R.; Sharifi-Rad, M.; Loizzo, M. R.; Ademiluyi, A. O.; et al. Biological activities of essential oils: From plant chemoecology to traditional healing systems. Molecules 2017, 22, 70–125.

33. Riyad, Y. M.; El-Kholany, E. A. Efficacy bioactive components of lavender (*Lavandula latifolia*) leaves as a natural antioxidant, antibacterial, and its uses as a cake preserving agent. Journal of Food and Dairy Science, Mansoura University 2020, 11(5), 113–120.

34. Beloued, A. Plantes médicinales d'Algérie. Office des Publications Universitaires, Alger, 2009; p. 134.

35. Nostro, A.; Scaffaro, R.; D'Arrigo, M.; Botta, L.; Filocamo, A.; Marino, A.; Bisignano, G. Development and characterization of essential oil component-based polymer films: A potential approach to reduce bacterial biofilm. Applied Microbiology and Biotechnology 2013, 97(21), 9515–9523.

36. Niloofar, K. J.; Ali, M. S. Antimicrobial effect of Melissa officinalis extracts on typical food-borne pathogens. International Journal of Advanced Life Sciences 2016, 9(1), 99–104.

37. Romeo, V.; Serena De Luca Piscopo, A.; Poiana, M. Antimicrobial effect of some essential oils. Journal of Essential Oil Research 2008, 2, 373–379.

38. Bağdat, R.; Coşge, B. The essential oil of lemon balm (*Melissa Officinalis* L.), its components and using fields. Journal of Faculty of Agriculture, OMU 2006, 21(1), 116–121.

39. Shakeri, A.; Sahebkar, A.; Javadi, B. *Melissa officinalis* L.: A review of its traditional uses, phytochemistry and pharmacology. Journal of Ethnopharmacology 2016, 188, 204–228.

40. Lee, S.; Cho, H.; Lee, K. G. Volatile compounds as markers of tofu (Soybean curd) freshness during storage. Journal of Agricultural and Food Chemistry 2014, 62(3), 772–779.

41. Della, R.; Mega, C.; Stefanus, S.; Maria, G. P. Application of Lemon Peel Essential Oil with Edible Coating Agent to Prolong Shelf Life of Tofu and Strawberry. AIP Conference 2017, 1803, 020037. https://doi.org/10.1063/1.4973164

42. Cagri, A.; Uspunol, Z.; Ryser, E. Antimicrobial edible films and coating. Journal of Food Protection 2004, 67, 833–848.

43. Bajpai, V. K.; Rahman, A.; Choi, U. K.; Youn, S. J.; Kang, S. C. Inhibitory parameters of the essential oil and various extracts of *Metasequoia glyptostroboides* Miki ex Hu to reduce food spoilage and food-borne pathogens. Food Chemistry 2007,105, 1061–1066.

44. Negi, P. S.; Chauhan, A. S.; Sadia, G. A.; Rohinishree, Y. S.; Ramteke, R. S. Antioxidant and antibacterial activities of various sea buckthorn (*Hippophae rhamnoides* L.) seed extracts. Food Chemistry 2004, 92, 119–124.

45. Riyad, Y. M.; Naeem, M. M. M.; Helmy, M. M.; Manal, A.; Sorour, M. A. Preparation of edible coating incorporated with lemon balm (*Melissa Officinalis* L.) for extending shelf life of tofu. International Journal of Scientific & Technology Research 2020, 9(8), 1–8.

Chapter 5

Major Preservation Technologies

Muhammad Shahbaz,[a] Jawad Ali,[a] Umar Farooq,[a] Shamas Murtaza,[a] Nighat Raza,[a] Atif Liaqat,[b] and Tahira Batool Qaisrani[c]

[a]*Department of Food Science and Technology, MNS-University of Agriculture, Multan, Pakistan*
[b]*Department of Food Science and Technology, Khawaja Fareed University of Engineering and Information Technology, Rahim Yar Khan, Pakistan*
[c]*Department of Agri Engineering and Technology, Ghazi University, Dera Ghazi Khan, Pakistan*
shahbaz.ft@mnsuam.edu.pk

Food spoilage is a major concern which can occur during harvesting, slaughtering, or manufacturing. There are different stages from procurement of raw materials to ultimate consumption by consumers where food spoilage can occur. At these stages, several control measures are in practice to ensure longer shelf life as well as high quality of the final product. These control measures aim to deliver safe food to the consumer. A broad range of physical, chemical, enzymatic, and microbiological reactions cause food spoilage which needs to be controlled to ensure the delivery of high-quality food

Natural Food Preservatives
Edited by Sourish Bhattacharya and Heba Hassan Abd-El Azim Salama

ISBN 978-981-4968-64-5 (Hardcover), 978-1-003-36776-5 (eBook)
www.jennystanford.com

products. Several preservation techniques which prevent food spoilage are freezing, dehydrating, salting, acidifying, pickling, vacuum packaging, the addition of preservatives, and controlled atmosphere packaging. Some preservation techniques prevent food spoilage by inactivating microorganisms such as ionizing radiation, sterilization, and pasteurization. In addition, there are some techniques that hinder the entrance of microbes into food products which include efficient packaging and aseptic processing. Non-thermal techniques of food preservation are a major trend which are being used or are under developmental stages and they preserve food by inactivating microbial cells. These techniques meet industry needs as they offer value-added products and are safer than thermal techniques (e.g., hydrostatic pressure, manothermosonication, antimicrobial enzymes, irradiation, pulsed electric field, ultrasound, pulsed X-ray, magnetic fields, pulsed light, hurdle technologies, and dense phase CO_2). Several preservation technologies are being developed that use a combination of different preservative aspects to ensure (1) minimum damage to the quality of the product, (2) controlled heating methods which supply moderate heat to food, (3) a combination of low- and high-temperature treatment that ensure satisfactory shelf life of products, (4) efficient packaging to maintain high quality for longer times, and (5) use of natural antimicrobial agents. Most of the preservation technologies interfere with the homeostatic mechanisms of microbes, thus hindering their multiplication and enhancing the shelf life as well as the quality of food commodities. This chapter summarizes the major trends in food preservation technologies.

5.1 Introduction

United Nations Economic and Social Affairs Department stated that in 2015 the population of the world was about 7.3 billion, which in 2030 is expected to reach about 8.5 billion, and in the year 2050 the total population will be about 9.7 billion and in 2100 it will be about 11.2 billion. To mitigate hunger, the global action suggests that to meet the needs of the swiftly expanding world population, we should enhance the production of food by about 70%. Nevertheless, the advancement and increase in farming practices are insufficient to solve the problem of increasing food demand. The industry can counter the increasing food demand of the world by evolution and worldwide execution of energy efficient, more advanced, and cost-effective technologies which also help to lower the impact on the environment and offer sustainability. According to the studies of the United Nations, nearly 2 billion tons of food is wasted, which is about 30–50% of the total food produced globally [1].

It is necessary to minimize food waste by spoilage because the world population is growing rapidly and to feed people, we should counter food waste. Owing to inadequate systems of food safety, the food industries lose millions of dollars via food waste, recalls of food products, reprocessing of finished products, and the resulting drop in sales. Foodborne ailments are no more restricted to developing areas of the world. As spoiled food is unfit for human consumption, it is wasted. Food spoilage can occur owing to unhygienic and improper handling, poor storage conditions, and inadequate temperature control, and it can be inhibited by using advanced and efficient techniques of food preservation [2].

Modern European program of food waste has specified that food waste due to the consumer is a key issue because in developed and developing nations most of the food waste occurs due to consumers throughout the whole food supply chain. A global food system is crucial which should be safe and minimize all types of food losses; it should also describe the ways to prevent and minimize food waste by consumers. Technologies of food preservation play a crucial role to reduce food waste and ensure the delivery of safe food to consumers [3].

In human life, food preservation has played a crucial role since prehistoric times. The earliest methods of food preservation used by humans were fermentation, salting, and sun drying; they were used to ensure food supply in times when there was no fresh food available. With the evolution of civilization, the demand for better quality and a huge quantity of processed food has increased, and it is further increasing day by day. Consequently, this increase in demand increases the interest of food preservation industries to deliver safe, nutritional, cost-effective, and satisfying food to consumers [4].

Furthermore, technologies of food preservation reduce the degradation of food and increase food utilization by implementing numerous novel and conventional methods. Novel methods of extraction, preservation, and processing possibly enhance production and promote a sustainable environment by minimizing the use of hazardous chemicals, production of hazardous material, wastewater prevention, and efficient use of fossil energy [5].

The food processing industry is occupied by heat treatment of foodstuffs, and it is termed a traditional method of food preservation. Oxidation in biological systems and foods causes numerous detrimental effects on food protection and stability and it also causes damage to human health. Hence, to protect human health and minimize food waste, several commercial, conventional, and novel technologies of food preservation or their combination should be used so that food items can be stored and used for longer times.

5.2 Thermal Technologies for Food Preservation

The most used and known procedure of food preservation to kill microorganisms and inactivate enzymes is the thermal treatment; it ensures safety to public health and provides longer shelf life for food products and makes them commercially reliable. Conditions of the process are easy to control during the treatment in this method [6]. In this method, thermal energy/heat is used to process foods where hot fluid is used to heat food to a particular temperature. Food is kept at a high temperature for a calculated time under which the characteristics of the product are not disturbed, known as processing time, and then the cold fluid is used to terminate thermal actions and cool down the product [4]. Pasteurization and sterilization are the significant conventional thermal techniques applied to preserve food for longer times.

5.2.1 Pasteurization

Pasteurization is a very crucial method of food preservation and is required to ensure food safety. It destroys almost all pathogenic microorganisms as well as other bacteria which cause food spoilage, and in doing so very less changes occur in the sensory properties and nutritional value of food. It uses mild heat treatment and is commonly utilized to increase the keeping quality of different foods; for example, in such food items that have low acid and high water activity, shelf life is increased up to a few days, while it is increased up to months in products having high acid and/or low water activity [6]

Mostly, the pH value of food products is used to specify the intensity of heat to which food will be subjected to enhance shelf life. In food items which contain pH more than 4.5, destruction of disease-causing bacteria is significant, while at a pH value below 4.5, inactivation of enzymes and destruction of spoilage microbes are significant. In pasteurization, the temperature ranges from 60°C to 75°C, which is applied for a specific time to preserve different food products such as fermented drinks, beer, fruit juices, and milk under atmospheric pressure. As pasteurized products show low shelf life and low stability, this method should be used in combination with another preservation technique to ensure high-quality food products [6].

5.2.2 Sterilization

Another thermal method which is used to preserve food is sterilization. In this method, intensive heat treatment, that is, above 100°C (115–130°C), is applied to foods to inactivate microorganisms. This method kills vegetative bacteria, yeast, molds, and spores. It enhances the shelf life of the product and makes it stable at ambient temperature. During storage, foods having pH 4.6 can be contaminated with *C. botulinum* spores, and so they should be subjected to sterilization for about 3 min at a temperature of 121°C to achieve a reduction of microorganisms by 12D [3].

In this process, firstly the product is heated to a high temperature of 110–125°C. Then the product is left at ambient temperature for a few minutes so that the temperature of the product equilibrates. Heat treated product should be kept at equilibrium temperature for a defined time period so that the value of sterilization can be ensured, and it is determined by the F_0 value (if the z value is 10°C and the temperature of sterilization is 121°C, then the time needed for sterilization for this temperature is demonstrated as the F_0 value) [7].

Sterilization is of two types; the first one is complete sterilization in which it is ensured that there is no viable microorganism in the product. In this type of sterilization, product quality is lost and the nutritional value of the product is reduced due to high temperature. The second type is commercial sterilization which kills all disease causing and food spoiling microbes that can deteriorate food quality under normal storage conditions. The main purpose of this type is to protect the quality of food and retain the nutritional characteristics of food [3].

5.3 Cooling and Freezing Technologies

5.3.1 Cooling

Cooling is the process in which food products can be stored at temperatures below 15°C and above freezing point while maintaining the liquid phase of water. The growth of microorganisms slows down during cooling, and it also delays the chemical reactions such as enzyme catalyzed oxidation/oxidative browning, color changes, dehydration, and post-slaughter and postharvest metabolic activities of animal tissues and plants, respectively. Commonly it is used in combination with other preservation technologies because the biochemical reactions and microbial and physicochemical activities continue under cooling conditions [4, 6].

Lowest and ideal temperature ranges for the growth and multiplication of four types of microbes are, respectively, thermophiles (30–40°C), (55–65°C); mesophiles (5–10°C), (30–40°C); psychrotrophs (0–5°C), (20–30°C); and psychrophiles (0–5°C), (12–18°C). Therefore, the growth of microorganisms can be retarded by cooling, and it inhibits microbial multiplication, but it does not confirm the inactivation of microorganisms. Hence, it should be used in combination with other preservation technologies [6].

5.3.2 Freezing

Although various new preservation methods are gaining importance and popularity, freezing technique is the best method to prolong the keeping quality of food products. It is a traditional and simple method of food preservation, and it has a major advantage that it can make food stable without losing the initial quality of food [4]. As water is transformed into ice during freezing, it changes the physical state of the substance. At freezing temperature, water changes into ice because heat is removed by the cooling phenomenon. During storage, the general temperature for freezing is around –18°C [8].

The process of freezing occurs in three steps. First is a precooling or chilling stage in which the product is cooled to its freezing point. In this stage, the removal of sensible heat occurs and it reduces the temperature of free water to simplify the crystallization of ice. The second is the phase transition stage in which latent heat of crystallization is moved away. In this stage, water is converted into ice during the process of crystallization. The freezing efficiency and quality of the frozen product are determined at this stage, so it is the most important step of freezing. The final stage is the tempering stage in which the product is cooled to the storage temperature [3, 9].

The quality of the frozen products can be enhanced by modulating the process of freezing as well as managing the preparation and storage conditions. In general, the process of fast freezing is preferred over the process of slow freezing, but longtime exposure of some products to extremely low temperatures can severely damage them. Contraction and volume expansion that occurs during ice formation can indicate this freeze cracking damage [10].

5.4 High-Pressure Technologies

Pressure is the perpendicular force applied on the surface of an object per unit area and the unit of pressure is pascal. Various kinds of pressure may be applied to an object: 1) isostatic pressure; it applies equal pressure in all directions of the product; 2) non-isostatic pressure, in which a pressure gradient is induced; and 3) dynamic pressure, which is the pressure of a fluid that results from its motion and high pressure is developed in a short period of time [11].

5.4.1 High Hydrostatic Pressure Processing

It is defined as an increase in isostatic pressure of a fluid owing to compression of the fluid and in relation to the food industry, the range of this pressure generally is from 100 to 700 MPa. Mainly, the HP equipment comprises two components, that is, a pressurizing unit and a cylindrical pressure vessel. There are two methods (direct and indirect methods) to build up pressure in the HP equipment. In the direct method, high pressure is delivered in the pressure chamber by a piston known as a pressure intensifier located in the vessel and it is done by reducing volume mechanically. In the indirect method, high pressure is induced by using a pump which is separate from the vessel; this pump delivers pressure into the pressure vessel by utilizing pressuring medium [12].

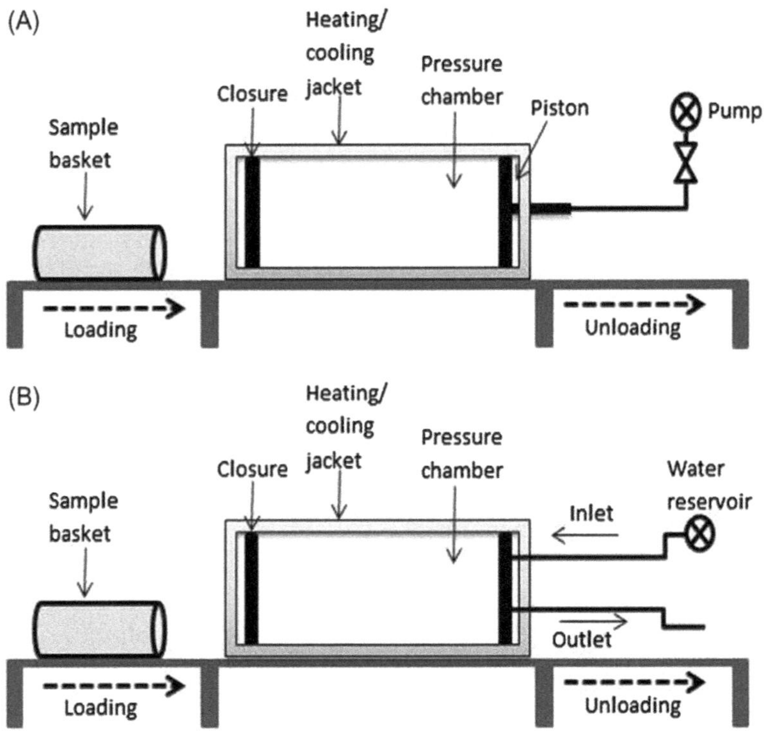

Figure 5.1 Pictorial depiction of (A) direct and (B) indirect high hydrostatic pressure systems [11].

The foodstuff, either liquid or solid, is packed in flexible containers such as bottles or bags in such a way that small headspace is left. The containers are then kept in the basket and shifted to the HP chamber which possesses a medium to transfer pressure. The vessel is closed by end closures and during pressurization, the end closures are kept in place by using a specialized yoke structure. The pump and piston compress the pressure transmitting medium, which increases the pressure in the pressure vessel. When the high-pressure treatment is completed, the closures are opened after decompressing the vessel and the sample basket is released from which pressure treated products can be collected [11]. The demonstration of full-operating HP equipment is shown in Fig. 5.1.

In the pressure chamber, pressure is transmitted to the food product by using pressure transmitting fluid (preferably water). Hence, such pressure transmitting fluids which are incompressible are much better to transmit pressure to foodstuffs and they also have a side benefit that the risk of explosion is considerably lower as compared to the gaseous medium [12]. HHP (high hydrostatic pressure) enhances the shelf life of food products including fruits and beverages by inactivating the microorganisms (Table 5.1).

Table 5.1 Impact of HHP on inactivation of microorganisms in fruit products

Food product	Targeted microorganism	Treatment framework	Log reduction of microbial load	Ref.
Apple jam	*Listeria monocytogenes*	20°C, 200 MPa for 5 min	2.9	[13]
Orange juice	*Escherichia coli*, *L. innocua*	241 MPa for 3 min	5	[14]
Apple juice	*E. coli* 29055	25°C, 400 MPa for 2 min	Less than 5	[15]
Apple juice	*Salmonella*, *E. coli*, *L. innocua*	25°C, 545 MPa for 1 min	5	[16]
Apricot, cherry juice	*S. enteritidis*, *E. coli* 0157:H7, *S. aureus*	30°C, 350 MPa for 5 min	Less than 5	[17]
Orange juice	*E. coli*	30°C, 550 MPa for 5 min	6	[18]

5.4.2 High Pressure Homogenization

In the homogenization process, the dynamic pressure is utilized which is produced by a mobile fluid. In this technology, the droplet size is reduced by using the special processing equipment called homogenizers. In 1899, Auguste Gaulin invented the first homogenizer which was used to stabilize a fat emulsion by breaking down larger fat globules into smaller ones (Patent number US753792A issued in 1904). At present, homogenization has gained much importance in the milk industry and is considered a standard industrial procedure. Therefore, microbial inactivation has become possible due to technological advancements in homogenization [19].

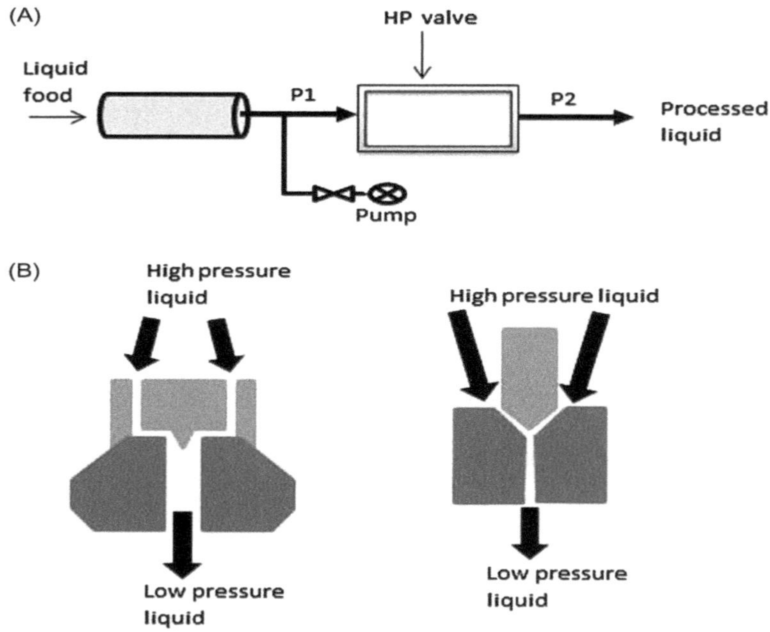

Figure 5.2 (A) Pictorial representation of HP valve and HP pump. (B) Pictorial demonstration of different HP valves [21].

Generally, in HPH (high pressure homogenization) a fluid is passed through a small orifice by applying high pressure which creates a pressure gradient between the outlet and the inlet of the orifice producing severe shear forces as well as a short-life heating phenomenon, turbulence, impact, and cavitation. The pressure of the homogenizer machine can be regulated according to our desire. The slow flowing fluid (46 m/s) is passed by the nozzle and homogenization valve and its pressure is significantly reduced due to the difference in the diameter of pipes; hence, the speed of flowing fluid is markedly increased, and it starts flowing at the speed of 200–250 m/s [20].

Initial application of pressure in HPH results in the rise of temperature to about 2–3°C/100 MPa owing to the heat of compression like high hydrostatic pressure (HHP) processing. Therefore, while passing through the homogenizing valve, this temperature rise is increased to about 14–18°C/100 MPa, which is because of shear forces, and mechanical input converts into heat. However, the temperature of the inlet fluid also influences the temperature rise, and it can raise temperature to about 17–21°C/100 MPa [21]. A pictorial demonstration of HP homogenization is shown in Fig. 5.2.

5.5 Electromagnetic Technologies

In this section, the mechanisms of action and basic principles of various advanced technologies based on electromagnetism such as electrolyzed water, ultraviolet light, radio frequency, microwaves, and electric field to preserve food items will be discussed. All the technologies based on electromagnetism are categorized into two categories, that is, non-thermal technologies and thermal technologies. In thermal technologies, foodborne pathogens are killed by using heat energy to preserve food, whereas microbial inactivation in non-thermal technologies is carried out by using other mechanisms instead of heat energy.

5.5.1 Electrolyzed Water

In recent times, it has been recognized that electrolyzed water (EW) is an important non-thermal sanitizer for the preservation of food. Two distinct types of water are produced by the process of electrolysis: (1) alkaline or reduced water having high pH and (2) oxidized or acidic water having low pH. Intense antimicrobial activity is shown by acidic electrolyzed water (AEW), but it is not evident yet whether this potential is due to pH, chlorine compounds, oxidation-reduction potential, or a combination of such factors [22].

It was assumed that the reduction in bacterial count occurs due to the acidic nature of electrolyzed water because hypochlorous acid makes the bacterial cells sensitive and destroys them by disturbing normal reactions and metabolism of nucleic acids, amino acid decarboxylation, and disruption of bacterial membranes [23].

A combination of alkaline electrolyzed water and acidic electrolyzed water was used to determine its sanitizing efficacy on freshly cut cilantro. Results demonstrated that in comparison to the individual treatment, this combination is much more effective in reducing the microbial load on a food product. In another study, fish quality was studied during cold storage by applying a combination of chitosan and acidic electrolyzed water. Results indicated that fish treated with this combination exhibit better texture, good sensory characteristics, and attractive color [24].

5.5.2 Pulsed Electric Field

Since the last century, the technology of pulsed electric field has gained much importance among researchers and attracted their interest, and in recent times its use on the industrial level has started. Research on pulsed electric field has been carried out to decontaminate foods, but biological cells are also affected by the pulsed electric field, so it appears to be applicable in various other fields of food processing including pretreatment prior to drying, canning, and freezing and it can be used to extract bioactive components from different foods [25].

In this technique, electric pulses are applied to products by using electrodes. The strength of pulses is in the range of 0.1–80 kV/cm, whereas the time of electric pulse generally ranges from μs up to ms. A capacitor bank is connected to a power supply of high voltage direct current to gather an enormous amount of energy which is eventually going to be applied to the food product as a high voltage electric pulse in the treatment chamber to enhance the shelf life of foods (different types of capacitor circuit can generate pulses of different shapes) [26].

Table 5.2 Impact of PEF on inactivation of microbes in fruit products

Food product	Targeted microorganism	Treatment framework	Log reduction of microbial load	Ref.
Cherry juice, peach nectar, apricot nectar	*Penicillium expansum*	21°C, 34 kV/cm, 163 μs	Complete inactivation of spores	[28]
Apple juice, grape juice	*E. coli*	55°C, 34 kV/cm, 7.68 μs	6.26.4	[29]
Orange juice	*S. typhimurium*	55°C, 90 kV/cm, 100 μs	5.9	[30]
Apple cider	*E. coli* O157:H7	42°C, 90 kV/cm, 20 μs	5.9	[31]
Orange juice	*Lactobacillus brevis*	54°C, 35 kV/cm, 2.5–5 μs	5.8	[32]
Strawberry juice	*E. coli* O157:H7	45°C, 18.6 kV/cm, 150 μs	3.09	[33]
Apple juice	*E. coli*	800 pulses/s, 36 kV/cm	6	[34]

In different foods, the inactivation of yeasts and various vegetative cells of bacteria has been achieved by using the technology of the pulsed electric field. In 1986, Zimmerman demonstrated that the application of electric pulses causes the breakdown of cell membrane which ultimately destroys the cell. This theory stated that microorganisms possess a membrane which may be considered a capacitor in which dielectric material is filled having a low dielectric constant. Consequently, at each membrane surface, free charges can be gathered, whereas when electric pulses are applied, it enhances the potential of membranes by increasing the load of those charges on the surfaces. Membrane thickness is reduced due to compression pressure which is generated by the attraction of opposite charges on the membrane surface. Hence, Zimmerman (1986) proposed that the electrocompression of membrane thickness is opposed by viscoelastic or elastic restoring force. When a high voltage electric pulse is applied, a breakdown of the membrane occurs due to a rapid increase in electric compressive forces and a decrease in membrane thickness as compared to elastic restoring compression forces [27].

The pulsed electric field can certainly kill the microorganisms and it retains the quality, fresh taste, nutritional profile, and flavors of the food due to low or ambient temperatures during processing [25]. Pulsed electric fields enhance the keeping quality of food including fruits and beverages by inactivating the microorganisms (Table 5.2).

5.5.3 Ohmic Heating

Ohmic heating is demonstrated as a procedure in which current is passed by such food products that exhibit an electrical resistance. The principle of the mechanism of ohmic heating is the discharge of electric energy as heat by using an electric conductor [35]. The amount of electric field generated depends on the operating parameters including intensity of applied voltage and time duration. Contrarily, the parameter of electrical conductivity is associated with the kind of foodstuff, and it can be measured by measuring the current transmitting through a unit area in per unit time [36].

Ohmic heating is a continuous in-line system which is used for sterilization and cooking of viscous pumpable liquid foods and mixtures of foods containing particulate foodstuffs. In ohmic heating, the inactivation of microorganisms is naturally thermal, but slightly non-thermal damage to the cell wall is also noticed owing to the existence of an electric field. The main advantage of ohmic heating is that it reduces the *D* value for the inactivation of microorganisms as compared to traditional methods of heating [12].

Besides temperature, the factors that greatly affect the electrical conductivity of various foods include the heating process, physical geometry, moisture mobility, and ionic content. So far, there are many applications of ohmic heating in numerous food processes including extraction, sterilization, pasteurization, blanching, and cooking, especially in liquid or particulate foods [37].

5.5.4 Microwaves

The use of microwaves is in practice for both domestic and industrial procedures for food manufacturing and processing. This process is very swift and needs less time to attain the process temperature. Hence, the comparison shows that sterilization and pasteurization processes are less efficient as compared to microwave heating. Microwave heating can overcome the limitations of traditional heating such as slow thermal diffusion [38].

Microwaves are electromagnetic in nature having wavelengths ranging from 1 to 0.001 m and their range of frequency lies between 300 MHz to 300 GHz. In food matrices, the polar molecules are influenced by microwave heating. Thus, the electromagnetic energy of microwaves is converted into thermal energy which is required for food sanitation and cooking. As volumetric heating requires very less time to achieve high temperatures, the commercial sterilization of food is done in less time and thermal damage to the desired compounds of food caused by long exposure to heat is reduced [12].

Microwave energy is transferred to the molecular level of food due to the molecular interaction of electromagnetic waves. Dissolved ions migrate due to conduction, thus causing molecular friction and dipole shift of polar solvents and increasing the internal temperature of food products. Thus, the main mechanisms of microwave heating are ionic polarization and dipole rotation. Heat is generated rapidly owing to this internal molecular friction [38].

5.5.5 Radio Frequency

Radiations of radio frequency lie in the array of electromagnetic field, which is at higher frequencies than microwaves. According to the European Telecommunications Standards Institute (ETSI) and Federal Communications Commission (FCC), most of the radio frequencies are used for telecommunications similar to microwaves except for radio waves of range 13.56, 27.12, and 40.68 MHz, which can be used for other purposes [39].

Molecular and ionic displacements are the two important mechanisms which occur during radio frequency heating. When a dielectric material is given an electric field, the ions move in opposite directions; that is, positive ions and negative ions move in opposite directions. The abovementioned procedure is demonstrated as the generation of heat by ionic displacement [35].

In molecular displacement, molecules having nonuniform charge distribution and native asymmetrical structure are involved. The electric field makes the molecules polarized and they try to realign themselves with regard to the electric field vector. The application of alternating electric field causes polar molecules to oscillate rapidly and generates heat within the food product owing to the frictional forces. The application of radio frequencies to sterilize deep food trays is more suitable as compared to microwaves because in various food types they possess five times higher penetration depth, whereas shallow food trays can be treated with microwaves [40].

5.5.6 UV Light

In the array of electromagnetic fields, ultraviolet (UV) radiations are found at higher frequencies as compared to radio frequency, microwaves, and infrared radiations. UV radiations are categorized into four categories, namely, vacuum UV, UV-A, UV-B, and UV-C, and their wavelength ranges from 100 to 400 nm. In the technologies of food preservation, the UV-C category is used because it is in the germicidal range and it can efficiently inactivate protozoa, viruses, fungi, and bacteria as compared to other categories [41].

Sources of ultraviolet light are differentiated by the level of power emission; low power levels generate continuous emission of UV radiations while high power levels generate pulsed emission of UV radiations. In the mechanism of microbial inactivation by using ultraviolet radiations, the most important parameter which influences the disinfection process is the power level. UV radiations create a modification in the structure of microbial cell deoxyribonucleic acid (DNA) which causes the destruction and inactivation of microbes [42].

So far the application of UV radiations proved to be effective in the fluid food type and surface application because their penetration depth is low, but the class UV-C is most effective against many microbes, and it may be implemented to make food items safe and is regarded as an efficient technology to preserve food for longer times and maintain its quality [43].

5.6 Sonic/Acoustic Technologies

5.6.1 Ultrasound

Food researchers and technologists are interested in the technology of ultrasound because it is capable to alter the physical properties of food products. As ultrasound technology provides heat to food media, it can be used as a substitute for the pasteurization of food [44].

The major factors which determine the efficiency of ultrasound treatment are treatment time, power intensity, and frequency. Commonly the ultrasound can be applied in a low energy regime (i.e., <1 W/m^2) or higher energy regime (i.e., >1 W/m^2) at the lower range of frequencies 20 kHz to $\approx$ 100 kHz (just above the human ear limit) or a higher range of frequencies such as above 100 kHz [45].

Various kinds of equipment can be used for the application of ultrasound to a food product such as vibrating systems, ultrasonic probes, and sonication baths; these systems can be installed in the continuous food processing line. In the case of solid foods, the process of sonication is done by placing the sonic probe and food sample in the processing equipment which acts as a reactor and is filled with the transfer media such as liquid or gas. In the case of liquid foods, the process of sonication can be done either in continuous processing lines or in batch reactors where radial vibrating plates or parallel vibrating plates are installed in the sonic chamber through which the liquid food is transported [44].

Process equipment can be of two types: (1) a vessel having ultrasonic probes which produce sonic waves in the system and (2) a radial vibrating system where vibrating motion is generated by the process unit itself. Cavitation bubbles are generated when high power ultrasound is applied to food media; they interact with neighboring cavitation bubbles and local media and ultimately crack up and generate an enormous amount of energy which leads to physical and chemical changes, thus causing inactivation of microbes [45].

In food processing, ultrasound can be used alone or in combination with other unit operations to enhance and intensify the process efficiency. Anyhow, it is mostly applied in combination with heat treatment, pressure, or both for better performance. Cavitation bubbles, rarefaction, and compression cause thinning of membranes in microbial cells and make them inactive, thus enhancing food safety [47]. Ultrasound treatments enhance the shelf life of food products including fruits and beverages by inactivating the microorganisms (Table 5.3).

Table 5.3 Impact of ultrasound processing on inactivation of microbes in fruit products

Food product	Targeted microorganism	Treatment framework	Log reduction of microbial load	Ref.
Apple juice	*Alicyclobacillus acidoterrestris*	300 W, 24 kHz, 60 min	80% inactivation of microbes	[48]
Apple cider	*L. monocytogenes*	600 W, 48 kHz, 25°C, 3 min	1 to 2	[49]
Apple cider	*E. coli*	600 W, 48 kHz, 25°C, 5 min	1 to 2	[49]
Apple cider	*L. monocytogenes*	750 W, 20 kHz, 60°C, 5 min	5	[50]
Orange juice	Aerobic mesophilic microbes	500 W, 20 kHz, 10°C, 8 min	1.38	[51]
Orange juice	Yeast and mold	500 W, 20 kHz, 10°C, 8 min	0.56	[51]
Carrot juice	*E. coli* K12	700–800 W, 19.3 kHz, 60°C, 1 min	2.5	[52]
Orange juice	Total mesophilic aerobic microbes	240 W, 500 kHz, 60°C, 15 min	3.4	[53]

5.6.2 High Hydrodynamic Pressure Shock Waves

In general, it relies on the application of shock waves up to 1 gigapascal (GPa) and it is produced due to dynamic force generated during the motion of fluids. In a fluid, shock waves travel faster compared to the speed of sound. Hence, they generate propagation of disturbance in the food media. The first high hydrodynamic pressure method consists of a tank which contains an explosive charge, and it is filled with the food medium to be treated and water. An explosion in water filled tank will immediately generate a shock wave which carries energy and apply pressure on the food media, and it will propagate quickly via the surrounding water in the tank [54].

5.7 Hurdle Concept

At present, consumers demand such food commodities which are free from physical damage and contain fewer chemical additives. Thus, in food preservation new opportunities for hurdle technology arise and gain importance. In food preservation, preservatives (e.g., antioxidants, sulfites, sorbate, and nitrite), targeted microbes (like lactic acid bacteria), acidity (pH), redox potential (E_h), temperature, and water activity (a_w) are the most important hurdle factors [6].

In hurdle technology, a combination of several inhibitory factors proves to be effective against microorganisms, but individually these factors are unable to prevent microbial growth. In food preservation, deteriorative changes such as protein denaturation, nonenzymatic browning, textural changes, flavor and aroma loss, and nutritional loss occur due to the application of severe preservation methods. Hence, a combination of two or more preservation methods enhances their efficiency and reduces the intensity of individual methods [3].

Frozen concentrated orange juice is an example of hurdle technology; it is conserved by using a combination of heat treatment and reduction of water activity by freezing and concentration. Sausages are pasteurized products and contain nitrites and nitrates (antimicrobial agents), but still they need to be kept at refrigeration temperatures. For the last two decades, hurdle technology has gained importance and is used extensively in food industries because conservation factors such as preservatives, competitive microorganisms, redox potential, water activity, acidity, and temperature are better understood [6].

5.8 Other Technologies

5.8.1 Membrane Filtration

The separation of particles from a liquid and its purification is done by using the technology of membrane filtration. A semipermeable membrane is used in this process and a solvent is passed through it. Mostly, electrical pumps are used in this system which are categorized into two classes: (a) centrifugal and (b) positive displacement [55].

The pore size of the membrane determines its permeability and it acts as a hurdle for larger particles, whereas the residue freely move by the membrane which produce a clean fluid. Semipermeable membranes can retain the compounds of high molecular weight such as microorganisms, colloids, and proteins, whereas compounds of smaller molecular weight can pass through the membranes such as vitamins, salts, water, and sugar [56].

The techniques which are commonly used for clarification of food media are nano, ultra, and microfiltration. In membrane filtration (MF), the semipermeable membrane having a pore size between 0.1 and 3 μm and pressure of about 100 psi ≈ 7 bar is used to separate compounds of high molecular weight from an aqueous solution. In ultrafiltration (UF), the semipermeable membrane having a smaller pore size of 0.02 μm and pressure of about 150 psi ≈ 10 bar is used to isolate solute from an aqueous solution. In nanofiltration (NF), semipermeable membranes having extremely small pore size (i.e., ≈1–10 nm) and pressure of about 450 psi ≈ 31 bar are used to carry out filtration [56].

It is noteworthy that the filtration technique for food products is most acceptable by the consumers because no preservatives are used in this technique. Additionally, the membrane filtration technique reduces the amount of waste produced. The major disadvantages of membrane filtration are cleaning or replacement of membranes and quick reduction of permeate flux which increases the operational and maintenance costs of the system [57].

5.8.2 Dense Phase CO_2

Pressurized carbon dioxide gas in supercritical fluid, gaseous, or liquid state is used in this technique. Beyond the critical temperature and pressure, the compound is called a "supercritical fluid." The density of such compounds is equivalent to that of a liquid while its viscosity is closer to that of a gas [58].

A conventional dense phase carbon dioxide (DPCD) system consists of a gas cylinder, immersible stirrer, data logger, pressure transducer, thermocouple, filter, needle valve, thermostatic controller, water container, pressure vessel, and pressure regulator. The food product is kept in touch with supercritical or pressurized carbon dioxide during DPCD treatment for a particular time duration and at a particular temperature either batch, semicontinuous, or continuous process [59].

The critical factors which influence the efficiency of DPCD are time, temperature, and pressure. Generally, the pressure used in the DPCD system ranges 7–40 MPa while the temperature of the system is considerably low (i.e., 30–50°C). Carbon dioxide in supercritical condition possess a higher ability to enter the cells because of enhanced cell membrane fluidity and CO_2 diffusivity, and hence temperature plays an important role to inactivate microorganisms caused by the DPCD system. Treatment time for batch and semicontinuous processes ranges from 120 to 140 min while for continuous processes the range of treatment time is 39 min [59]. A schematic demonstration of the DPCD system is shown in Fig. 5.3.

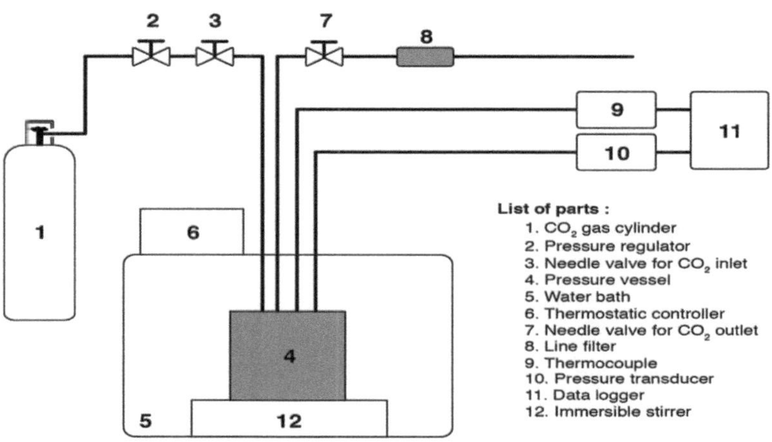

Figure 5.3 Pictorial representation of a DPCD system [59].

5.9 Conclusion

High-quality and safe foods can be delivered by emerging preservation technologies. Many problems that restrict the extensive implementation of these technologies have been recognized such as the deficiency of proper regulations and the huge investment required for the installation of systems. The use of hurdle technology can overcome the detrimental effects of other technologies on the nutritional, textural, and sensory qualities of different food products. The combination of different technologies improves the sensory and organoleptic properties of food products and ensures the provision of safe foods. Advanced technologies of food processing such as HPP, ohmic heating, EW, food irradiation, and PEF offer high-quality products, improved productivity, enhanced reliability, decreased emissions, and lower energy consumption. The combination of different techniques enhances the shelf life of food products, reduces the treatment effects on food products, minimizes the use of sanitizers, and reduces the expenditure on preservation treatment. In food industries, a combination of different technologies has been investigated, but further investigative studies are required to demonstrate the most efficient combinations of various treatments to increase the overall quality of food items and reduce the severity of the individual unit process.

References

1. FAO. 2015. Global initiative on food loss and waste reduction. http://www.fao.org/3/ai4068e.pdf.
2. Gustavson, J., Cederberg, C., Sonesson, U., van Otterdijk, R. and Meybeck, A. (2011). Global food losses and food waste: Extent, causes, and prevention. *Food and Agriculture Organization of the United Nations (FAO)*.
3. Tavman, S., Otles, S., Glaue, S. and Gogus, N. (2019). Food preservation technologies. *Saving Food*, 117-140.
4. Karel, M. and Lund, D. B. (2003). *Physical Principles of Food Preservation: Revised and Expanded*, vol. 129. CRC Press.
5. Chemat, F., Rombaut, N., Meullemiestre, A., Turk, M., Perino, S., Fabiano-Tixier, A. S. and Abert-Vian, M. (2017). Review of green food processing

techniques. Preservation, transformation, and extraction. *Innov. Food Sci. Emerg. Technol.* **41**, 357-377.

6. Augusto, P. E., Soares, B. M. and Castanha, N. (2018). Conventional technologies of food preservation. *Innov. Tech. for Food Preservation,* 3-23.
7. Ramesh, M. N. (2004). Canning and sterilization of foods. In *Handbook of Food Preservation.* CRC Press. pp. 603-642.
8. Cheng, L., Sun, D. W., Zhu, Z. and Zhang, Z. (2017). Emerging techniques for assisting and accelerating food freezing processes: A review of recent research progresses. *Crit. Rev. Food Sci. Nutr.* **57(4)**, 769-781.
9. Wu, X. F., Zhang, M., Adhikari, B. and Sun, J. (2017). Recent developments in novel freezing and thawing technologies applied to foods. *Crit. Rev. Food Sci. Nutr.* **57(17)**, 3620-3631.
10. Li, B. and Sun, D. W. (2002). Novel methods for rapid freezing and thawing of foods—a review. *J. Food Eng.* **54(3)**, 175-182.
11. Balasubramaniam, V. M., Martinez-Monteagudo, S. I. and Gupta, R. (2015). Principles and application of high pressure–based technologies in the food industry. *Annu. Rev. Food Sci. Technol.* **6**, 435-462.
12. Barba, F. J., Ahrné, L., Xanthakis, E., Landerslev, M. G. and Orlien, V. (2018). Innovative technologies for food preservation. *Innov. Tech. for Food Preservation,* 25-51.
13. Préstamo, G., Sanz, P. D., Fonberg-Broczek, M. and Arroyo, G. (1999). High pressure response of fruit jams contaminated with Listeria monocytogenes. *Lett. Appl. Microbiol.* **28(4)**, 313-316.
14. Guerrero-Beltran, J. A., Barbosa-Canovas, G. V. and Welti-Chanes, J. (2011). High hydrostatic pressure effect on natural microflora, Saccharomyces cerevisiae, Escherichia coli, and Listeria Innocua in navel orange juice. *Int. J. Food Eng.* **7(1)**.
15. Ramaswamy, H. S., Riahi, E. and Idziak, E. (2003). High-pressure destruction kinetics of E. coli (29055) in apple juice. *J. Food Sci.* **68(5)**, 1750-1756.
16. Avure Technologies. www.avure.com/archive/documents/.../juice-and-beverage-data.pdf.
17. Bayındırlı, A., Alpas, H., Bozoğlu, F. and Hızal, M. (2006). Efficiency of high pressure treatment on inactivation of pathogenic microorganisms and enzymes in apple, orange, apricot and sour cherry juices. *Food Control,* **17(1)**, 52-58.
18. Linton, M., McClements, J. M. J. and Patterson, M. F. (1999). Inactivation of Escherichia coli O157: H7 in orange juice using a combination of high pressure and mild heat. *J. Food Prot.* **62(3)**, 277-279.

19. Barba, F. J., Terefe, N. S., Buckow, R., Knorr, D. and Orlien, V. (2015). New opportunities and perspectives of high pressure treatment to improve health and safety attributes of foods. A review. *Food Res. Int.* **77**, 725-742.
20. Floury, J., Bellettre, J., Legrand, J. and Desrumaux, A. (2004). Analysis of a new type of high pressure homogeniser. A study of the flow pattern. *Chem. Eng. Sci.* **59(4)**, 843-853.
21. Dumay, E., Chevalier-Lucia, D., Picart-Palmade, L., Benzaria, A., Gràcia-Julià, A. and Blayo, C. (2013). Technological aspects and potential applications of (ultra) high-pressure homogenisation. *Trends Food Sci. Technol.* **31(1)**, 13-26.
22. Khan, I., Tango, C. N., Miskeen, S., Lee, B. H. and Oh, D. H. (2017). Hurdle technology: A novel approach for enhanced food quality and safety—A review. *Food Control* **73**, 1426-1444.
23. Rahman, S. M. E., Park, J. H., Wang, J. and Oh, D. H. (2012). Stability of low concentration electrolyzed water and its sanitization potential against foodborne pathogens. *J. Food Eng.* **113(4)**, 548-553.
24. Hao, J., Li, H., Wan, Y. and Liu, H. (2015). Combined effect of acidic electrolyzed water (AcEW) and alkaline electrolyzed water (AlEW) on the microbial reduction of fresh-cut cilantro. *Food Control* **50**, 699-704.
25. Zimmermann, U. (1986). Electrical breakdown, electropermeabilization and electrofusion. *Rev. Physiol. Biochem. Pharmacol.* **105**, 175-256.
26. Puértolas, E., Koubaa, M. and Barba, F. J. (2016). An overview of the impact of electrotechnologies for the recovery of oil and high-value compounds from vegetable oil industry: Energy and economic cost implications. *Food Res. Int.* **80**, 19-26.
27. Barbosa-Cánovas, G. V., Pothakamury, U. R., Gongora-Nieto, M. M. and Swanson, B. G. (1999). Preservation of Foods with Pulsed Electric Fields. Elsevier.
28. Evrendilek, G. A., Tok, F. M., Soylu, E. M. and Soylu, S. (2008). Inactivation of Penicillum expansum in sour cherry juice, peach and apricot nectars by pulsed electric fields. *Food Microbiol.* **25(5)**, 662-667.
29. Heinz, V., Toepfl, S. and Knorr, D. (2003). Impact of temperature on lethality and energy efficiency of apple juice pasteurization by pulsed electric fields treatment. *Innov. Food Sci. Emerg Technol.* **4(2)**, 167-175.
30. Liang, Z., Mittal, G. S. and Griffiths, M. W. (2002). Inactivation of Salmonella Typhimurium in orange juice containing antimicrobial agents by pulsed electric field. *J. Food Prot.* **65(7)**, 1081-1087.

31. Iu, J., Mittal, G. S. and Griffiths, M. W. (2001). Reduction in levels of Escherichia coli O157: H7 in apple cider by pulsed electric fields. *J. Food Prot.* **64(7)**, 964-969.

32. Elez-Martinez, P., Escolà-Hernández, J., Soliva-Fortuny, R. C. and Martín-Belloso, O. (2005). Inactivation of Lactobacillus brevis in orange juice by high-intensity pulsed electric fields. *Food Microbiol.* **22(4)**, 311-319.

33. Gurtler, J. B., Bailey, R. B., Geveke, D. J. and Zhang, H. Q. (2011). Pulsed electric field inactivation of E. coli O157: H7 and non-pathogenic surrogate E. coli in strawberry juice as influenced by sodium benzoate, potassium sorbate, and citric acid. *Food Control* **22(10)**, 1689-1694.

34. Charles-Rodríguez, A. V., Nevárez-Moorillón, G. V., Zhang, Q. H. and Ortega-Rivas, E. (2007). Comparison of thermal processing and pulsed electric fields treatment in pasteurization of apple juice. *Food Bioprod. Process.* **85(2)**, 93-97.

35. Barba, F. J., Orlien, V., Mota, M. J., Lopes, R. P., Pereira, S. A. and Saraiva, J. A. (2016). Implementation of emerging technologies. *Innov. Strategies in the Food Ind.* 117-148.

36. Goullieux, A. and Pain, J. P. (2005). Ohmic Heating. London: Elsevier Ed.

37. Knirsch, M. C., Dos Santos, C. A., de Oliveira Soares, A. A. M. and Penna, T. C. V. (2010). Ohmic heating–a review. *Trends Food Sci. Technol.* **21(9)**, 436-441.

38. Ahmed, J. and Ramaswamy, H. S. (2004). Microwave pasteurization and sterilization of foods. In *Handbook of Food Preservation.* CRC Press. pp. 713-732.

39. Pereira, R. N. and Vicente, A. A. (2010). Environmental impact of novel thermal and non-thermal technologies in food processing. *Food Res. Int.* **43(7)**, 1936-1943.

40. Wang, Y., Wig, T. D., Tang, J. and Hallberg, L. M. (2003). Dielectric properties of foods relevant to RF and microwave pasteurization and sterilization. *J. Food Eng.* **57(3)**, 257-268.

41. Gómez-López, V. M., Koutchma, T. and Linden, K. (2012). Ultraviolet and pulsed light processing of fluid foods. *Novel Thermal and Non-Thermal Technologies for Fluid Foods*, pp. 185-223.

42. Sinha, R. P. and Häder, D. P. (2002). UV-induced DNA damage and repair: A review. *Photochem. Photobiol. Sci.* **1(4)**, 225-236.

43. Guerrero-Beltr· n, J. A. and Barbosa-C· novas, G. V. (2004). Advantages and limitations on processing foods by UV light. *Food Sci. Technol. Int.* **10(3)**, 137-147.

44. Zinoviadou, K. G., Galanakis, C. M., Brnčić, M., Grimi, N., Boussetta, N., Mota, M. J. and Barba, F. J. (2015). Fruit juice sonication: Implications on food safety and physicochemical and nutritional properties. *Food Res. Int.* **77**, 743-752.

45. Knorr, D., Zenker, M., Heinz, V. and Lee, D. U. (2004). Applications and potential of ultrasonics in food processing. *Trends Food Sci. Technol.* **15(5)**, 261-266.

46. Rastogi, N. K. (2011). Opportunities and challenges in application of ultrasound in food processing. *Crit. Rev. Food Sci. Nutr.* **51(8)**, 705-722.

47. Chemat, F. and Khan, M. K. (2011). Applications of ultrasound in food technology: processing, preservation and extraction. *Ultrason. Sonochem.* **18(4)**, 813-835.

48. Yuan, Y., Hu, Y., Yue, T., Chen, T. and Lo, Y. M. (2009). Effect of ultrasonic treatments on thermoacidophilic Alicyclobacillus acidoterrestris in apple juice. *J. Food Process. Preserv.* **33(3)**, 370-383.

49. Rodgers, S. L. and Ryser, E. T. (2004). Reduction of microbial pathogens during apple cider production using sodium hypochlorite, copper ion, and sonication. *J. Food Prot.* **67(4)**, 766-771.

50. Baumann, A. R., Martin, S. E. and Feng, H. (2005). Power ultrasound treatment of Listeria monocytogenes in apple cider. *J. Food Prot.* **68(11)**, 2333-2340.

51. Gómez-López, V. M., Orsolani, L., Martínez-Yépez, A. and Tapia, M. S. (2010). Microbiological and sensory quality of sonicated calcium-added orange juice. *LWT-Food Sci. Technol.* **43(5)**, 808-813.

52. Zenker, M., Heinz, V. and Knorr, D. (2003). Application of ultrasound-assisted thermal processing for preservation and quality retention of liquid foods. *J. Food Prot.* **66(9)**, 1642-1649.

53. Valero, M., Recrosio, N., Saura, D., Muñoz, N., Martí, N. and Lizama, V. (2007). Effects of ultrasonic treatments in orange juice processing. *J. Food Eng.* **80(2)**, 509-516.

54. Solomon, M. B., Liu, M. N., Patel, J. R., Bowker, B. C. and Sharma, M. (2006). Hydrodynamic pressure processing to improve meat quality and safety. *Technologies for Meat Processing.* Taylor and Francis. pp. 219-245.

55. Baker, R. W. (2012). *Membrane Technology and Applications.* John Wiley & Sons.

56. Chen, Y., Yu, L. J. and Rupasinghe, H. V. (2013). Effect of thermal and non-thermal pasteurisation on the microbial inactivation and phenolic

degradation in fruit juice: A mini review. *J. Sci. Food Agric.* **93(5)**, 981-986.

57. Rodriguez-Gonzalez, O., Buckow, R., Koutchma, T. and Balasubramaniam, V. M. (2015). Energy requirements for alternative food processing technologies—principles, assumptions, and evaluation of efficiency. *Compr. Rev. Food Sci. Food Saf.*, **14(5)**, 536-554.

58. Balaban, M. O. and Duong, T. (2014). Dense phase carbon dioxide research: Current focus and directions. *Agric. Agric. Sci. Procedia*, **2**, 2-9.

59. Damar, S. and Balaban, M. O. (2006). Review of dense phase CO_2 technology: Microbial and enzyme inactivation, and effects on food quality. *J. Food Sci.* **71(1)**, 1-11.

Chapter 6

Diverse Content of Preservatives for Industrial Food Products

Fiona Sarah DSouza[a] and Anupama Shrivastav[b]
[a]*Department of Microbiology, Parul Institute of Applied Sciences, Parul University, Vadodara, Gujarat, India*
[b]*Faculty of Life Health and Allied Sciences, Institute for Technology and Management - ITM Vocational University, Vadodara, Gujarat, India*
anupamashrivastav@gmail.com

Preservation of food as an agent is really important to ensure that manufactured food products remain safe and nondestructive. In this chapter, we will learn about different modes of both chemical-derived and biological-derived preservatives and the response mechanisms of these compounds in microorganisms for the concern to the food industry. Lactic acid bacteria (LAB) and metabolites obtained from preservatives are being considered major components of biopreservative as they lack a lot of benefits on human health which has been recognized by GRAS (generally recognized as safe) which belongs to American Food and Drug Administration (FDA). Some LAB produce bacterial proteins themselves which are known as bacteriocins; they target specific

Natural Food Preservatives
Edited by Sourish Bhattacharya and Heba Hassan Abd-El Azim Salama

ISBN 978-981-4968-64-5 (Hardcover), 978-1-003-36776-5 (eBook)
www.jennystanford.com

bacteria to inhibit their growth. Bacteriocins perform their property by forming pores in the targeted bacteria and by applying thermal stress systemically and provide a great amount of pH activity which results in inhibition of further activity. Starter culture plays a vital role by providing a kick-start for the production of fermented products and having the coculture relationship with the products to be formed. Like bacteriocins, nisin, being a part similar to that which is generated by *Lactococcus lactis*, has the property of antimicrobial activity against Gram-positive bacteria. In spite of the differences, the chemical or natural preservatives become part of the food production with having a relationship directly or indirectly during stages of its rectifying, packaging, or even storage. Preservation of food products is fundamentally necessary to extend the limited shelf life of food. Nowadays there are a diverse amount of chemicals for food preservation, but these needs to be appreciated by the responsible food safety authority present. The preservation of food has the aim of having responsibility of food safety, preventing or inhibiting foodborne diseases worldwide, and preventing or declining the spoilage of food or degradation that makes the food unapproved in the world.

6.1 Introduction

A higher level of preservatives explains the diversity of antimicrobial and antioxidant actions of naturally found compounds to prevent bacterial and fungal spoilage, where the growth of microbial substances may cause a deterioration of the food product for a great amount of spoilage. There is currently occurring a large interest in substituting synthetic food preservatives and obtaining biological food preservatives which increase the preservation of industrial obtained food products. However, especially in Europe, a particular group of the food industry is keen to exploit natural ingredients and additives that make sure to protect against food quality deterioration and prevent the spoilage of food products. The safety and quality of food products are paramount for consumer health and for the betterment of individual's aspects. Preservatives are most prominently required to have a good maintenance of quality, long shelf life, and the guarantee of the safety of food products.

Therefore, most preservation techniques both chemically and biologically acquired aim to dominate all forms of spoilage. Synthetically produced chemical preservatives have an essential role in food preservation. Chemicals which can be utilized industrially such as sodium benzoate, butylated hydroxyanisole, sodium nitrite, butylated hydroxytoluene, and potassium sorbate have been most frequently used in food products. The food industry uses a variety of diverse chemicals to prevent microbial growth that triggers food deterioration to occur [1]. Their application comes beside the use of high-temperature processing equipment which makes sure to help in inclining the prolonged life of the food by essentially disturbing the activities of pathogenic and spoilage occurring microorganisms; however, the organoleptic properties of the food may have more destructive effects to healthy life [2].

6.2 Preservation of Food and Its Role

Preservation plays an immense role in saving the food shelf life and increasing the shelf life hand in hand. The definition clearly states that the preservation of foods includes preventing the growth of microorganisms such as yeasts and other microbes and slowing the rate of the fats and lipids which result in rancidity. The pH level of food products changes according to the ingredients involved in them and further processing the final product to take place. Thus the entire process is taken under consideration where the authorities lie the input in the food product.

Food products can get contaminated by bacteria, fungi, and other microorganisms present in the environment. Most of these microorganisms can cause disagreeable reactions that may depreciate various results such as flavor, odour, color, and different visual properties of food as well as cause the food to result in spoilage. Thus, fortification of foods is mandatory during their procedures, storage, and distribution to provide them the desired shelf life [3].

The spoilage of food is currently an appropriate concern because some microorganisms can theoretically cause foodborne illness and there is great economic loss associated with food loss. This has stimulated the food industry to look for substitute preservatives that could enhance the safety and quality of foods [4].

Microorganisms can contaminate food in numerous ways, including at the farm level (irrigation water, laborers in field, insects, fecal contamination from wild animals), as well as other sources like handling by workers, transport duration of the product in vehicles, processing instruments, and cross-contamination occurring from other foods [5].

The demand for minimally administered, easily equipped, and ready-to-eat fresh foods, the globalization of food production worldwide, and distribution from centralized processing to other places throughout pose major challenges to food safety and superiority [3].

It has been predicted that approximately more than 48 million cases of foodborne illness occur in the United States every 12 months, and it has been estimated that 32% of the obtainable food supply at the trade and consumer levels in 2010 was not consumed at a great amount, which is equivalent to a commercial loss of almost a fortune of USD 161.6 billion, which was the biggest loss until the current situation [6].

Control of food putrefaction and pathogenic bacteria is mainly accomplished by chemical control and several preservation techniques which all together (heat treatment, managing of salt concentration, measuring their pH level, and drying) have been formerly used in the food industry for a long time. Nevertheless, the usage of synthetic chemicals in the industry is limited due to undesirable characteristics including carcinogenicity (cancer), acute toxicity (increased rate of the toxin), teratogenicity (cause damage to the developing fetus), and slow degradation periods, which could lead to environmental problems such as pollution [1].

The negative public awareness of industrially amalgamated food antimicrobials has generated interest in the use of more naturally occurring combinations, receiving a good deal of attention in widespread searches, for the potential use of natural food additives that could preserve a broad continuum of antioxidant and antimicrobial activities by either decreasing the development of antibiotic resistance by pathogenic microorganisms or reinforcement of immune cells in humans, while possessing the ability to recover the quality and shelf life of consumable foods [7].

Antimicrobials are chemical mixtures that are naturally present in or added to foods, food packaging, and food processing environments to inhibit microbial growth or kill microorganisms present in the food. The primary functions of antimicrobials in food are (1) to slowdown natural spoilage processes (food preservation) and (2) to prevent/rheostat growth of microorganisms, including pathogenic microorganisms (food safety) [1].

An ideal naturally occurring antimicrobial should (1) be operative at low concentrations in its natural form, (2) be cost-effective, (3) cause no sensual changes to the product, (4) inhibit a wide range of pathogenic and spoilage by microorganisms, and (5) be nontoxic. Compounds derived from natural regular acquired sources have the potential to be used for food safety due to their antimicrobial properties against a wide range of food causing pathogens known for foodborne diseases. They can be obtained from different sources including plants (e.g., essential oils of various leaves currently present worldwide such as basil, clove, thyme, cinnamon, oregano, and rosemary), microbial (e.g., nisin and bacteriocin), animals (e.g., enzymes), organic acids (e.g., sorbic, propionic, citric acid), natural polymers (e.g., chitosan), algae, and fungi [4, 8, 9].

Other possible antibacterial products are derivatives from fruits and vegetables which could be a cheap source of natural antimicrobials [10]. The use of natural or controlled microbiota to extend the shelf life and enrich the safety of foods is defined as biopreservation or biocontrol [11]. Lactic acid bacteria (LAB) occur naturally in many food systems and have a long history of safe expenditure in fermented foods, and thus they were classified as GRAS (generally recognized as safe) and have a great probability of being used in biopreservation. On this basis, they have been approved as a food additive with GRAS status in over 80 countries globally.

6.3 Chemical Preservatives

Food preservation using chemicals are basically known as chemical preservatives which further can be described as the chemical additives that are supplementary to foods specifically to prevent the spoilage or decomposition of a food and are safe for human ingestion. According to the FDA (Food and Drug Administration), food additives are directly and indirectly added to food products to inhibit many sources by becoming an element or otherwise affecting the individualities of any food [12]. There are several factors that influence the activity of chemical preservation, such as pH concentration of preservatives, types, and the number of microbial flora existing in the food, the temperature of the foods which are kept for preservation, and the chemical and physical properties of foods. The main resolution of using chemical preservatives in foods is to inhibit the growth and activities of spoilage causing microorganisms in the outcome.

6.3.1 Organic Acids

Organic acids are chemicals that have been primarily used in food industries for preservation for ages. The disadvantageous effect of organic acids on microorganisms makes it appropriate for utilization in preservation. Pathogens are primarily responsible for causing foodborne diseases and are exterminated by acid shock during food processing [13]. The most frequently used organic acids in food industries as preservatives are lactic acid, propionic acid, sorbic acid, acetic acid, benzoic acid, and citric acid.

- Acetic acid: Acetic acid has an exceptional function at a higher pH when associated with other organic acids. Acetates nevertheless are ordinarily effective on bacteria but also have a fungistatic effect up to a particular extent on yeasts and molds [14]. This is the purpose behind the consumption of acetic acids in mostly all packed food items. Dehydroacetic acid, monochloroacetic acid, peracetic acid, and sodium diacetate are the universally used acetic acid as forms in preservation. Vinegar is a mutual form of acetic acid that is developed in the preservation of pickles. Sodium diacetate is particularly used to prevent the putrefaction and adulteration of malt syrup, which is the raw material for beer preparation and also cheese spreads, ketchup, and mayonnaise along with the wrappers which are cast off to wrap butter [15].
- Benzoic acid: These weak organic acids are exclusively used in jellies, carbonated soft drinks, jams, and fruit juices. Sodium benzoate is the most extensively used form of benzoic acid, which is mostly operative at lower pH and thus can be used

in sauces, fruit juices, pickles, and soft drinks [16]. Propyl paraben and methyl paraben are the two major ester forms of benzoic acid highly used in food preservation industries.

- Propionic acid: Propionic acid has resilient fungicide movement. However, they are also effective against bacteria that are available in the food products.
- Sorbic acid: This strong antifungal chemical is able to hinder the growth of yeasts and molds. The effect of sorbic acid is strongly efficient for slaying yeasts, as well as the bacteria responsible for biofilm formation. In experimentation in the preservation of tangerine sherbet base with sorbic acid, the chemical was able to prove its effectiveness without any change of flavor [17].

6.3.2 Nitrites and Nitrates

Combinations of nitrites with nitrates are used essentially to preserve the packed meats and meat products. Nitrate is auxiliary added as potassium or sodium salt. Nitrite has a strong inhibitory effect on a wide range of spoilage causing bacteria including *Flavobacterium*, *Micrococcus*, *Pseudomonas*, *Escherichia coli*, *Salmonella*, and *Lactobacillus*. The primary everyday reputation of nitrates is the fact that they have a sturdy sporicidal activity, which leads them to be useful against the various spore forming strains of *Clostridium botulinum*, a bacterium that habitually causes spoilage of meats and results in life-threatening food poisoning in humans. *N*-nitrosamines are utilized in many food products for the anticipation of microbial growth [18]. The interaction of nitrites with other constituents at appropriate pH yields various compounds like nitrous acid, which acts as an active antimicrobial agent. The intervention of nitrites with the iron and enzymes encompassing the sulfhydryl groups in bacterial cells further inactivates them. Such activities are uniquely responsible for the antibacterial activity of nitrites and nitrates. There are four ways of bacterial inhibition that are commonly targeted during preservation techniques: effects on the genetic system, damage to cell wall or cell membrane, enzymes and protein inactivation, and binding of essential nutrients. Nitrites and nitrates are pragmatic to affect bacterial cells. The activity of nitrite and nitrosyl compounds on pyruvate:ferredoxin reductase in *C. botulinum* results in the formation of iron-sulfur-nitric oxide complex [19].

6.3.3 Sulfur Dioxide and Sulfites

Sulfur dioxide is widely recycled in food processing industries for the purpose of sterilization of equipment or appliances. The broad application of sulfur dioxide ranges from prevention of microbial contamination to powerful antioxidants and enzyme inhibitors. Sulfur dioxide is primarily used as an antioxidant that results in the inhibition of the enzyme browning reaction in grape juices recycled for wine preparation. Apart from juices, sulfur dioxide is also used as a preservative for a wide range of syrups, sauces, purees, jams, jellies as well as fermented beverages. The peroxiding effect of sulfur dioxide aids in the preparation of watermelon rind, maraschino cherries, and Zucca melon and it is well documented [35]. In commercial marketplaces, sulfites are used in the preservation of dried fruits and vegetables, pickles, and soft drinks. In the presence of water, sulfur dioxide produces sulfurous acid at low pH, which is a resourceful antibacterial compound. Sulfurous acid can constrain a wide group of spoilage causing bacteria. At high pH, sulfur dioxide produces sulfite ions, which are correspondingly important as chemical preservatives [20].

6.3.4 Ethylene Oxide and Propylene Oxide

Unlike most of the other types of chemical preservatives, oxides are non-bleaching compounds that can be used as gases. Ethylene oxide is able to kill all types of microorganisms that are present in any particular food sample. The microorganisms that are inhibited and killed by ethylene oxide also include fungus. Propylene oxide acts in the same manner as ethylene oxide but with subordinate reactivity when compared to ethylene oxide. These two disinfecting agents are customarily used for the preservation purpose of dried products, such as dried fruits, eggs, and grains [25]. Epoxides act by alkylation of the biomolecules, largely proteins and enzymes. The hydrogen atom present in the amino group and carboxylic group is reversibly replaced by ethylene oxide. As a result, the proteins become nonfunctional. Ethylene oxide also replaces the sulfhydryl groups present in many bacterial enzymes and thus put out of action to them. These gases are also used in the spice industry [21, 22].

6.3.5 Sugars and Salts

These compounds shrink the water activity of food products and thus have an argumentative effect on microbial growth. Sodium chloride is employed in brine and is further directly applied to the foods. The apparatus followed by sodium chloride to inhibit bacterial growth is well assumed. It intensifies the osmotic pressure, which ultimately leads to plasmolysis of the bacterial cell, but a suitable percentage of sodium chloride has to be used to prevent bacterial growth in foods. Sodium chloride works as one of the dehydrating agents that withdraw water from the microbial cells. Another mechanism of sodium chloride activity in the prevention of microbial growth is ionization where it ionizes to form chloride ions, which are destructive to cells. When added in high concentration, sodium chloride reduces the oxygen content of food products and hence inhibits the development of aerobic microbes. Sodium chloride has also been found to interfere with proteolytic enzymes concealed by spoilage causing bacteria and fungi further avoiding the food spoilage. Sugars present worldwide, such as fructose, lactose, and sucrose, are also modern chemical preservatives. It convinces osmotic imbalance by reducing the water content of the food and thus prevents the growth of spoilage causing microbes [23]. The food products that contain a high level of sugars, such as fruit syrups, condensed milk, or jams, are not usually ruined because of the added glucose or sucrose. In the challenging experiments with salt and sugar preservation, the mutual effect of the solutes was checked on apples which are highly available, which showed that it inclines the rate of osmosis when used in amalgamation [24].

6.3.6 Alcohol

Alcohol is a protein denaturing agent that is effective at a concentration of approximately 60–90%. It may also turn by precipitation of necessary bacterial enzymes. Alcohol also worsens bacterial cell membranes by dissolving the membrane lipids. It is also a resourceful dehydrating agent and thus destroys the bacterial cells because of the absence of water. The great amount of alcoholic content increasing on a daily basis in beers, wines, or other alcoholic beverage makes them unsuitable for microbial spoilage. Glycol and glycerol are sugar alcohols, which are also extensively used to impede the growth of fungi and airborne bacteria that may contaminate the foods at the time of packaging. Glycerol has high competence in alleviating protein, as well as preventing the accumulation of protein [25].

6.3.7 Spices and Condiments

Spices and condiments may not have an effective antimicrobial activity but their addition to the foods has been found to prevent putrefaction for a long period of time. They are mostly used in households and they do not have any commercial solicitations. The impulsive oil present in the mustards is found to always have appropriate effects against strains of *Saccharomyces* [7]. The mustard oil is effective against a collection of microorganisms due to the manifestation of allyl isothiocyanate in it, which is a strong antimicrobial agent [3]. Other condiments, such as cloves and cinnamon, consist of a molecule called cinnamic aldehyde, which has a bacteriostatic activity in the food products. Ginger and black pepper also cover both antifungal and antibacterial activities [26]. Other plant products produces and utilized in food processing techniques also have the highest amount of bacteriostatic action. Garlic contains an organic compound called acrolein, which has an inhibitory effect on *Bacillus subtilis* and *E. coli*, both of which are elaborated in food spoilage. Spices may also inhibit the growth of fungal particles by inhibiting the formation of asexual fungal spores.

6.3.8 Antibiotics

Most of the antibiotics are produced to preserve raw foods with high protein concentrations, such as raw meat (chicken, turkey, red meat), fish, and poultry products. Antibiotics are prohibited for claiming in any of the processed or packaged foods. Various types of antibiotics are used for the prevention of the destruction of raw meat and other poultry products, which can be listed as chloramphenicol, bacitracin, oxytetracycline, streptomycin, polymyxin, neomycin, natamycin, and so on. Among these drugs mentioned, bacitracin and natamycin are highly used to prevent fungal progress in enveloped fruit juices and are also used in cheese in some cases. Natamycin is largely used to inhibit microbial contamination in meat and fish products. *L. lactis* produces an antibiotic called nisin, which is used in cheese as well as other tinned food to prevent microbial contamination [27]. The debate about antibiotic utilization in food and its side effects on humans has involved the FDA to assign the antibiotics in three criteria, namely, low, medium, and high, according to the effect they have on humans and also whether the antibiotic has an appropriate activity for preservation [28]. Apart from these preservatives involved, other chemicals are also highly involved, specifically phosphoric acid, boric acid, and hydrogen peroxide, which have persuasive antimicrobial activity (Table 6.1).

Table 6.1 Chemical preservatives utilized by different food industries worldwide and their percentages of usage

Preservatives	Percentages used worldwide
Sucrose	1%
Sodium chloride	2% and 4%
Acetic acid	0.1% and 0.3%
Ascorbic acid	0.015% and 0.055%
Benzoic acid	0.05% and 0.1%
Potassium metabisulfite	0.05% and 0.25%
Sodium benzoate	0.05% and 0.1%
Sodium metabisulfite	0.05% and 0.25%
Sodium sulfite	0.05% and 0.25%
Sodium nitrate	0.02% and 0.05%
Sorbic acid	0.025% and 0.1%
Tartaric acid	1%
Citric acid	0.02% and 0.035%
Calcium propionate	0.1% and 0.4%

6.4 Microbial Preservatives

Preservation ordinarily involves precluding the growth of bacteria, fungi (such as yeasts), and other microorganisms present whole universal, as well as impeding the oxidation of fats which cause rancidity. A numeral amount of procedures were invented for prevention can be used that can either totally preclude, deferment, or otherwise diminish food spoilage. Sustaining or generating texture, nutritional value, and flavor is an important characteristic of food preservation.

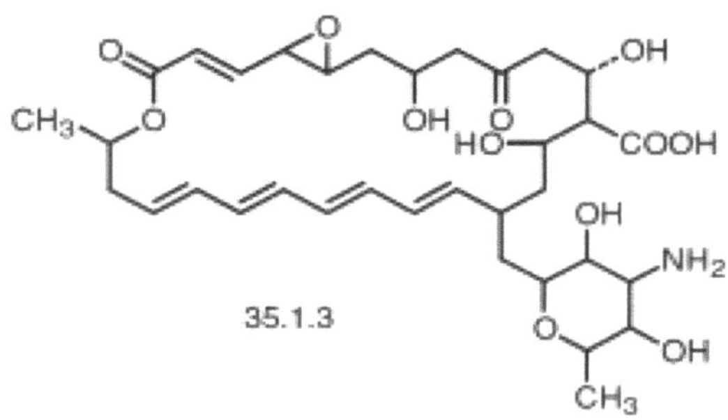

Figure 6.1 Structure of natamycin.

6.4.1 Natamycin

Natamycin is one of the antifungal agents that have its place to the group comprehending of polyene macrolide antimycotics and is generated by fermentation using Streptomyces natalensis [29] (Fig. 6.1). Natamycin has a high similarity to ergosterol and binds permanently to the ergosterol in the fungal cell membrane. This agitates the permeability of the cell membrane entirely, which informs to the speedy outflow of essential ions and small peptides, thereby initiating cell lysis [30]. Natamycin was very lately recognized and was approved in 1982 in the United States to be applied in cheese in demand to prevent mold growth, with focused concentrations of 20 mg/L in the finished food product. The MIC (minimum inhibitory concentrations) of natamycin against nearly all foodborne causing fungi is a smaller amount than 25 ppm; however, the solubility of natamycin in aqueous food systems found in the water system is around 41 ppm. Practically, under suitable hygienic conditions, this concentration of dissolved natamycin was validated to be sufficient to prevent fungal growth without encumbering the bacterial fermentation in cheese [31].

6.4.2 Nisin

Nisin is one of the most microbial usage which is a cationic polypeptide with 34 amino acids and was characterized as a class Ia bacteriocin or sometimes also lantibiotic (Fig. 6.2) [32]. It is produced by certain strains of *L. lactis* [33]. Nisin was also one of the approved compounds in the United States in 1988 for inhibiting the growth of *C. botulinum* spores and the establishment of toxins in pasteurized cheese suppers and has consequently been applied to other foods. In Europe, it was further added to the food additive list in the early 1930s (also known as E234) and is the only bacteriocin that has been recommended by the World Health Organization, which is one of the organizations followed by civilians worldwide, to be used as a food preservative [11, 34]. Nisin acts at the cytoplasmic membrane, interrelating electrostatically with membrane phospholipids to yield transient, nonselective pores. Nisin displays a high affinity for lipid II, the widespread exporter of cell wall peptidoglycan components [35]. The rapid efflux of amino acids, ions, and adenosine triphosphate concluded the pores results in the collapse of the transmembrane proton reason force and cell death [36]. Nisin is most effective against Gram-positive vegetative bacteria and spores of *Clostridium* spp. and *Bacillus* strains. The usage of nisin in combination with a chelating agent such as ethylenediamine tetra-acetic acid (EDTA) may intensify the effectiveness. Furthermore, the effect of nisin can also be heightened by using treatments such as osmotic shock, exposure to sublethal heat, and freezing, because these managements make the cell wall of Gram-negative microorganisms more penetrable and therefore more susceptible to nisin [37]. Predominantly, the cast-off of bacteriocin-producing LAB (or their more or less disinfected bacteriocins) has also acknowledged increment of interests. These peptides show strong antimicrobial commotion against closely related bacteria. Microgard is a commercially accessible fermented milk product containing antimicrobial metabolites which are an impending inhibitor for Gram-negative bacteria such as *Salmonella* spp., *Pseudomonas* spp., and *Yersinia* spp. And almost all Gram-positive organisms are present (except *Listeria innocua*) in dairy products [38]. Food defensive cultures are foods cultured with microorganisms that can illustrate an inhibitory effect against other microorganisms as an outcome of their competition for nutrients; production of peptides or other antagonistic compounds such as hydrogen peroxide, organic acids, propionates, and enzymes; or a combination of these entirely [39]. They are highly applied in the food industry to encompass the shelf life of food and to prevent the growth and development of foodborne pathogens.

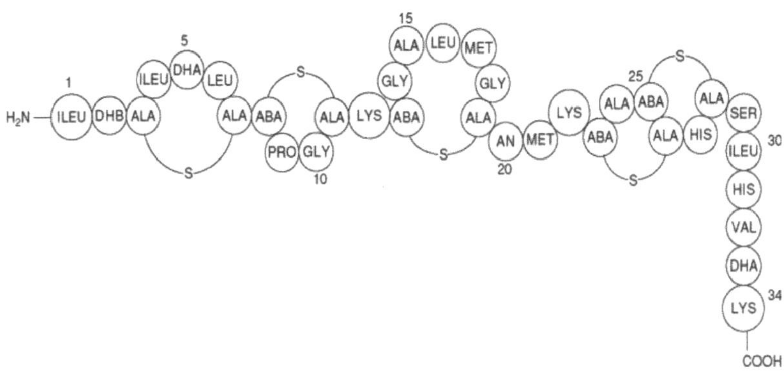

Figure 6.2 Structure of nisin. Reprinted from Ref. [55], Copyright 2006, with permission from Elsevier.

6.4.3 Defensive Cultures Used for Food Products

Many species of microbes have been used as protective cultures, including not only LAB, for example, enterococci, lactobacilli, streptococci, Lactococci spp., and bifidobacteria but also *E. coli* strain as well as the species of *Bacillus* spp., yeasts such as *Saccharomyces*, and molds such as *Aspergillus*. Nevertheless, the most corporate protective cultures for human consumption belong to *Lactobacillus* spp. (e.g., *L. casei*, *L. rhamnosus*, *L. johnsonii*, *L. acidophilus*, and *L. reuteri* and species of *Bifidobacterium* (e.g., *B. breve*, *B. bifidum*, and *B. longum*) [40]. The utilization of food protective cultures has attracted attention, exclusively in the meat, dairy, and seafood industries, for the reason that they are safe for consumption (GRAS) and naturally dictate the microbiota of many foods [41]. There are several industrial appetizers such as Bovamine Meat Culture (NPC, United States), SafePro (CHR Hansen, Denmark), and HOLDBAC (Danisco, Denmark) that have been industrialized to switch *L. monocytogenes* for the meat industry [2].

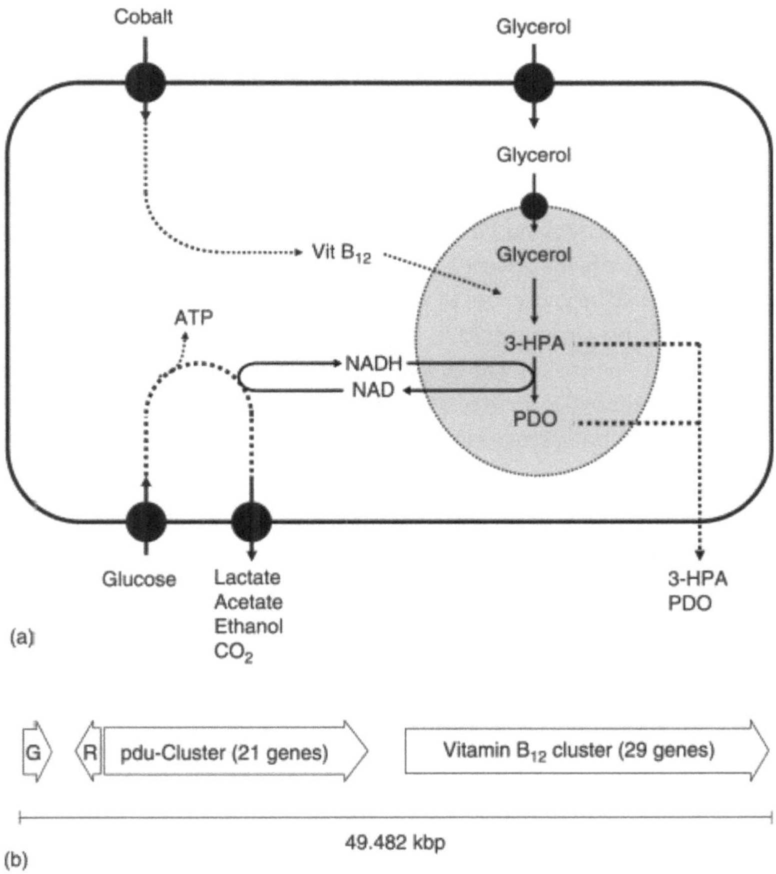

Figure 6.3 Reuterin components used in *L. reutrii*. Reprinted with permission from Ref. [57].

6.4.4 Reuterin

Reuterin (also known as β-hydroxypropionaldehyde) is a molecule with antimicrobial commotion toward a broad continuum of spoilage organisms and foodborne pathogens. Its lipolytic and proteolytic enzymes, resistance to heat, high solubility in water, and stability over a wide range of pH values make reuterin an ideal biopreservative for foods (Fig. 6.3) [42].

Arqués et al. [20] validated the antimicrobial activity of reuterin in amalgamation with different bacteriocins from LAB against foodborne pathogens which can be contaminants in milk at a refrigerated temperature known to be psychrophiles. A synergistic conclusion of reuterin in combination with nisin was also observed on *S. aureus* and *L. monocytogenes* [43, 44]. The procedure of reuterin to control both Gram-negative and Gram-positive pathogens as well as microorganisms has been scrutinized in dairy and also in meat products. These discoveries have shown that reuterin has a complex antimicrobial activity on Gram-negative bacteria which has a lot more than on Gram-positive pathogenic bacteria. Even though the mechanism of antimicrobial activity is not yet vibrant, it is thought that the aldehyde form of reuterin is the bioactive agent which causes an oxidative stress response by transforming thiol groups in proteins [45].

6.4.5 Bacteriocins

Bacteriocins are known as ribosomal synthesized antimicrobial proteins containing peptides that are produced by bacteria having antibacterial commotion headed for closely related species, such as *Staphylococcus aureus*, Ecolab, *L. monocytogenes*, or even different strains of the similar species. These are heterogeneous groups that go through significant posttranslational modifications, for example, formylation and glycosylation of the N-terminal residue. These modifications make a great distinction between one bacteriocin and another in respect of many aspects of activities taken place as antimicrobial one. The cells that generate bacteriocins are insusceptible due to the production of "immune protein" coded by imm gene. These proteins can be prompted in a cell by exposure to certain both physical and chemical agents. Bacteriocins illustrate synergistic special effects when cast off in amalgamation with organic compounds, heat, and also pressure as a part of food packaging. These days, among Gram-positive bacteria, especially, lactobacilli enlarged responsiveness due to their property of producing bacteriocins as a product. Food industries are predominantly anxious about spores due to their thermal resistance and chemical properties in their quiescent states. Bacteriocins were characterized into four different groups depending on their individualities, such as molecular weight, producer organisms, and mode of action. According to several research, solicitation of bacteriocins in food preservation confirmations numerous reimbursements to the food industry along with to consumers. For instance, (I) condensed financial losses owing to food putrefaction, (II) compact risk of foodborne ailment that is relocated throughout the entire food chain available, (III) additional microbial fortification under incompatible storage conditions, (IV) expanded food shelf life, (V) curtailed use of synthetic food preservatives, and (VI) loads of industries and consumers, such as fresh tasting, safe, and so on [46–50].

6.4.5.1 Types of bacteriocins

Bacteriocins are classified into four classes by Klaenhammer in 1993 constructed on chemical structures which postulate the differences as well as similarities witnessed during the discovery of each molecule of bacteriocins.

Class I group bacteriocins have infrequent amino acids, such as β-methyl-lanthionine/lantibiotics, dehydrobutyrine, and dehydroalanine, and have small peptides, such as nisin. Nisin is of two alternatives, that is, nisin A and nisin Z, showing the same molecular outline except for only one amino modification.

Class II bacteriocins are small unmodified peptides and heat labeled with molecular size less than 10 kDa or more and divided into two classes based on pore-forming peptides and antilisterial peptides. Class IIa consists of pediocin-like lantibiotics, such as pediocin PA1 and leucocin A, showing antilisterial activity like one of the kind. Class IIb bacteriocins essential two or more original peptides in order to stimulate antimicrobial activity by pore-forming complexes. These peptides can be independently active by showing synergetic conclusion when acting in composed form.

Class III bacteriocins have a single thiol (-SH) residue and are, consequently, referred to as thiolbiotics. This group consists of heat-labile proteins with a molecular mass of more than 30 kDa or so. Class III bacteriocins are also termed as one of the bacteriolysins which are endopeptidases and tempt lysis of the cell wall of the target microorganisms through their enzymatic action.

Class IV compounds do not comprehend cysteine residues in the bacteriocin complex. The nonexistence is arguable as it includes peptide bacteriocins that are not yet purified properly.

Later these four of the groups were differentiated and classified as classes into two forms, which were Class I lantibiotics and Class II nonlantibiotics or other peptides without lanthionine.

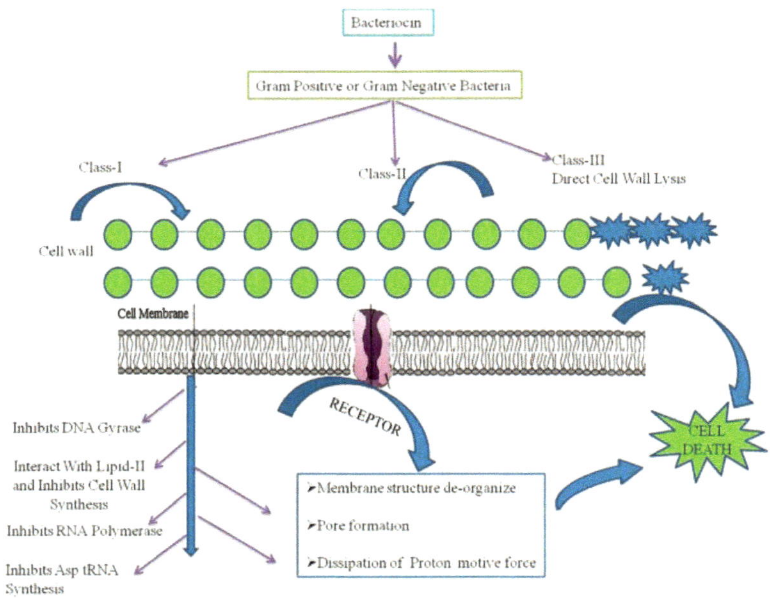

Figure 6.4 Mechanisms of bacteriocins and their classes. Reprinted from Ref. [56], Copyright 2017, with permission from Elsevier.

Class I has modified groups that include lantibiotics along with linardins, sactibiotics, thiopeptides, glycocins, and so on. Class II contains not modified or cyclic groups that include IIa, IIb, IIc, IId, and IIe peptides obtaining groups involved. In 2006, bacteriocins were distributed into three foremost classes, namely, Class I, Class II, and Class III, and these three classes were further differ from each other with respect to genetic and biochemical characteristics involved in each of them [51–54].

6.4.5.2 Biosynthesis of bacteriocins

As we studied previously about bacteriocins being a ribosomal synthesized antimicrobial peptides possessing prepeptide at the N-terminal region as a leader peptide and attached to a peptide having a C-terminal region on the other end. The leader peptide also safeguards bacteriocin producer by preserving lantibiotics in its inactive state until it is traded out of the cell.

Immeasurable steps involved in the biosynthetic pathway of lantibiotics are (1) splicing of the leader peptide, (2) posttranslational modification, (3) biosynthesis of the peptide, and (4) transport of the processed peptide outside the cytoplasmic membrane [55–58].

In the case of Class I bacteriocins, nisin and subtilin, they are external auto regulator in the signal transduction process, while in the case of Class II lantibiotics, the beginning of transcription is prompted by the induction factor that is a stable and a cationic hydrophobic protein. ABC transporter cleaves IF which is present in forerunner peptide during transportation of mature bacteriocin from cells. In Class II bacteriocins, prepeptide is composed of N-terminal region, which is vastly preserved and has a proteolytic property processing site. A two-element system is habitually involved in the instruction of bacteriocin biosynthesis of both lantibiotics and nonlantibiotics. They further include a large amount of histidine protein kinase (HPK) and membrane-bound cytoplasmic regulator (CR). Bacteriocin is added superficially to the fermentation medium to enhance and induce the production of bacteriocin at an industrial level by automatic phosphorylation process of highly conserved histidine residue by HPK in the intracellular domain.

This immunity avoids the producer bacterium from its own bacteriocins by attributing ownself to the outer surface of the cytoplasmic membrane and irregular bacteriocin interaction with the producer cell surface.

6.5 Conclusions

In the food industry, preservatives of bacteriological origin have long been proposed as a solution to the difficulties of foodborne infections and food spoilage. Bacteriocin producing lactic acid bacteria have been developed as great substitutions for chemicals and antibiotics in the arena of food technology. Some bacteriocins validated antimicrobial recreational activities against food spoilage pathogens. In the modern era of antibiotic resistance, LAB and their metabolites, namely, bacteriocins, may be the only preparation for hindrance of microbial growth. These include nisin and chitosan in the form of dynamic packaging. The effectiveness of preservatives of microbial origin in foods is dictated by environmental factors. For the application of each individual preservative, there is a need to conclude more precisely more operative conditions. Nevertheless, the combined application of many other machine hurdle technologies still remains unexplored. With the introduction of novel technologies for integrating bacteriocins more accurately and bacteriocin producing strains into the food milieu and packing material, which is auspicious and practicable at the industrial scale, the demand for bacteriocin in food preservation is anticipated to escalate in the approaching years.

References

1. Jayant, D. and Halami, P.M., 2020. Industrial perspective of food preservatives from microbial origin. Available at https://doi.org/10.1016/B978-0-444-64309-4.00011-8.
2. Chatterjee, A. and Abraham, J., 2018. Microbial contamination, prevention, and early detection in food industry. Available at https://doi.org/10.1016/B978-0-12-811515-2.00002-0.
3. Barberis, S., Quiroga, H.G., Barcia, C., Talia, J.M., Debattista, N., 2018. Natural food preservatives against microorganisms. Available at https://doi.org/10.1016/B978-0-12-814956-0.00020-2.
4. Davidson, P.M., Zivanovic, S., 2003. The use of natural antimicrobials. In: Zeuthen, P., Bogh-Sorensen, L. (Eds.), Food Preservation Techniques. Woodhead Publishing Ltd., Cambridge.
5. Weber, G., Steenson, L., Delves-Broughton, J., 2008. Antimicrobial fermentate technology. In: Havkin-Frenkel, D. (Ed.), Proceedings of the

Second International Symposium on Natural Preservatives in Food, Feed, and Cosmetics. Acta Hort, International Society for Horticultural Science, Amsterdam.

6. Stark, J., Roller, S., 2003. Natamycin: an effective fungicide for food and beverages. In: Roller, S. (Ed.), Natural Antimicrobials for the Minimal Processing of Foods. Woodhead Publishing Ltd., Cambridge.
7. Klaenhammer, T.R., 1993. Genetics of bacteriocins produced by lactic acid bacteria. FEMS Microbiol. Rev. 12.
8. Kupke, T., Gotz, F., 1996. Expression, purification, and characterization of EpiC, an enzyme involved in the biosynthesis of the lantibiotic epidermin, and sequence analysis of Staphylococcus epidermidis epi C mutants. J. Bacteriol. 178.
9. Sobrino-Lopez, A., Martin-Belloso, O., 2008. Enhancing the lethal effect of high-intensity pulsed electric field in milk by antimicrobial compounds as combined hurdles. J. Dairy Sci. 91.
10. Wiedemann, I., Breukink, E., Van Kraaij, C., Kuipers, O.P., Bierbaum, G., De Kruijff, B., 2001. Specific binding of nisin to the peptidoglycan precursor lipid II combines pore formation and inhibition of cell wall biosynthesis for potent antibiotic activity. J. Biol. Chem. 276.
11. Adams, M., 2003. Nisin in multifactorial food preservation. In: Roller, S. (Ed.), Natural Antimicrobials for the Minimal Processing of Foods. Elsevier, Boca Raton.
12. Gálvez, A., Abriouel, H., López, R.L., Omar, N.B., 2007. Bacteriocin-based strategies for food biopreservation. Int. J. Food Microbiol. 120.
13. Von Staszewski, M., Jagus, R.J., 2008. Natural antimicrobials: effect of Microgard and nisin against Listeria innocua in liquid cheese whey. Int. Dairy J. 18.
14. Melero, B., Vinuesa, R., Diez, A., Jaime, I., Rovira, J., 2013. Application of protective cultures against Listeria monocytogenes and Campylobacter jejuni in chicken products packaged under modified atmosphere. Poult. Sci. 92.
15. Gibson, G.R., 2008. Prebiotics as gut microflora management tools. J. Clin. Gastroenterol. 42.
16. Fliss, I., Hammami, R., Lay, C.L., 2011. Biological control of human digestive microbiota using antimicrobial cultures and bacteriocin. In: Lacroix, C. (Ed.), Protective Cultures, Antimicrobial Metabolites and Bacteriophages for Food and Beverage Biopreservation. Woodhead Publishing, Cambridge.

17. Castellano, P., Belfiore, C., Fadda, S., Vignolo, G., 2008. A review of bacteriocinogenic lactic acid bacteria used as bioprotective cultures in fresh meat produced in Argentina. Meat Sci. 79.
18. Pilet, M.F., 2011. Applications of protective cultures, bacteriocins, and bacteriophages in fresh seafood, and seafood products. In: Lacroix, C. (Ed.), Protective Cultures, Antimicrobial Metabolites and Bacteriophages for Food and Beverage Biopreservation. Woodhead Publishing, Cambridge.
19. Gyawali, R., Adkins, A., Minor, R.C., Ibrahim, S.A., 2014. Behavior and changes in cell morphology of Escherichia coli O157:H7 in liquid medium and skim milk in the presence of caffeine. CyTA–J. Food. 12.
20. Arqués, J.L., Rodríguez, E., Núñez, M., Medina, M., 2011. Combined effect of reuterin and lactic acid bacteria bacteriocins on the inactivation of food-borne pathogens in milk. Food Control 22 (3–4).
21. Bian, L., Molan, A.L., Maddox, I., Shu, Q., 2011. Antimicrobial activity of Lactobacillus reuteri DPC16 supernatants against selected food borne pathogens. World J. Microbiol. Biotechnol. 27.
22. Stevens, M., Vollenweider, S., Lacroix, C., 2011. Potential of reuterin produced by Lactobacillus reuteri as broad spectrum preservative in food. In: Lacroix, C. (Ed.), Protective Cultures, Antimicrobial Metabolites and Bacteriophages for Food and Beverage Biopreservation. Woodhead Publishing, Oxford.
23. Langa, S., Landete, J.M., Martín-Cabrejas, I., Rodríguez, E., Arques, J.L., Medina, M., 2013. In situ reuterin production by Lactobacillus reuteri in dairy products. Food Control 33.
24. Davidson, P.M., Critzer, F.J., Taylor, T.M., 2013. Naturally occurring antimicrobials for minimally processed foods. Annu. Rev. Food Sci. Technol. 4.
25. Gupta, R.K., 2007. Food Preservation. Food and Industrial Microbiology. National Science Digital Library.
26. Ricke, S.C., 2003. Perspectives on the use of organic acids and short chain fatty acids as antimicrobials. Poultry Sci. 82 (4).
27. Gurtler, J.B., Doyle, M.P., Kornacki, J.L. (eds.), 2014. The Microbiological Safety of Low Water Activity Foods and Spices. Springer, New York.
28. Silva, M.M., Lidon, F., 2016. Food preservatives: an overview on applications and side effects. Emirates J. Food Agric. 28 (6).
29. Levine, A.S., Fellers, C.R., 1940. Action of acetic acid on food spoilage microörganisms. J. Bacteriol. 39 (5), 499.

30. Sivasankar, B., 2002. Food Processing and Preservation. PHI Learning Pvt. Ltd., New Delhi.

31. Wibbertmann, A., Kielhorn, J., Koennecker, G., Mangelsdorf, I., Melber, C., 2002. Concise International Chemical Assessment Document 26. Benzoic Acid and Sodium Benzoate. World Health Organization, Geneva.

32 Patrick, R., Atkins, C.D., 1954. Effectiveness of sorbic acid as a preservative for tangerine sherbet base. Florida Agric. Exp. Station J. Ser. 292.

33. Wolff, I.A., Wasserman, A.E., 1972. Nitrates, nitrites and nitrosamines. Science 177 (4043), 15.

34. Cammack, R., Joannou, C.L., Cui, X.Y., Martinez, C.T., Maraj, S.R., Hughes, M.N., 1999. Nitrite and nitrosyl compounds in food preservation. Biochim. Biophys. Acta Bioenergetics 1411 (2).

35. Mrak, E.M., Stewart, G.F., 1954. Advances in Food Research. Vol. I. Advances in Food Research. Vol. V. No. 167.

36. Vally, H., Misso, N.L., Madan, V., 2009. Clinical effects of sulphite additives. Clin. Exp. Allergy 39 (11).

37. Gurtler, J.B., Doyle, M.P., Kornacki, J.L. (Eds.), 2014. The Microbiological Safety of Low Water Activity Foods and Spices. Springer, New York.

38. Daniels C., 2015. How salts and sugars work to preserve foods. Livestrong (June 21). Available at: http://www.livestrong.com/article/432144-how-salts-sugars-work-to-preserve-foods/.

39. Sereno, A.M., Moreira, R., Martinez, E., 2001. Mass transfer coefficients during osmotic dehydration of apple in single and combined aqueous solutions of sugar and salt. J. Food Eng. 47 (1).

40. Vagenende, V., Yap, M.G., Trout, B.L., 2009. Mechanisms of protein stabilization and prevention of protein aggregation by glycerol. Biochemistry 48 (46).

41. Frazier, W.C., 1993. Food microbiology. Micribiologia de los Alimentos (No. QR115. F718 1993).

42. Lucera, A., Costa, C., Conte, A., Del Nobile, M.A., 2014. Food applications of natural antimicrobial compounds. Antimicrobial compounds from natural sources. Front. Microbiol. 3, 287.

43. Adedeji, T.O., Ade-Omowaye, B.I.O., 2013. The preservative effects of two local Nigerian spices on the shelf life of fried bean cake snacks. J. Nutr. Food Sci. 3, 2.

44. Gould, G.W., 1995. Biodeterioration of foods and an overview of preservation in the food and dairy industries. Int. Biodeterior. Biodegrad. 36 (3).

45. Landers, T.F., Cohen, B., Wittum, T.E., Larson, E.L., 2012. A review of antibiotic use in food animals: perspective, policy, and potential. Public Health Rep. 127.

46. Landers, T.F., Cohen, B., Wittum, T.E., Larson, E.L., 2012. A review of antibiotic use in food animals: perspective, policy, and potential. Public Health Rep. 127.

47. Tajkarimi, M.M., Ibrahima, S.A., Cliverb, D.O., 2010. Antimicrobial herb and spice compounds in food. Food Control 21, 1199–1218. Talia, J., Debattista, N., Pappano, N., 2011.

48. Davidson, P.M., Bozkurt Cekmer, H., Monu, E., Techathuvanan, C., 2014. The use of natural antimicrobials in food: an overview. In: Taylor, T.M. (Ed.), Handbook of Natural Antimicrobials for Food Safety and Quality. Woodhead Publishing/Elsevier, Cambridge.

49. Scallan, E., Hoekstra, R.M., Angulo, F.J., Tauxe, R.V., Widdowson, M.A., Roy, S.L., 2011. Foodborne illness acquired in the United States major pathogens. Emerg. Infect. Dis. 17.

50. Buzby, J.C., Wells, H.F., Hyman, J., 2014. The Estimated Amount, Value, and Calories of Postharvest Food Losses at the Retail and Consumer Levels in the United States. Economic Information Bulletin. Department of Agriculture, Economic Research Service, USA.

51. Ramesh, C., Pattar, M.G., 2010. Antimicrobial properties, antioxidant activity and bioactive compounds from six wild edible mushrooms of Western Ghats of Karnataka, India. Pharm. Res. 2.

52. Bhagavathy, S., Sumathi, P., Jancy Sherene Bell, I., 2011. Green algae Chlorococcum humicola—a new source of bioactive compounds with antimicrobial activity. Asian. Pac. J. Trop. Biomed. (1).

53. Baydar, N.G., 2006. Organic acids, tocopherols and phenolic compositions of some Turkish grape cultivars. Chem. Nat. Compd. 42.

54. Sobrino-Lopez, A., Martin-Belloso, O., 2008. Enhancing the lethal effect of high-intensity pulsed electric field in milk by antimicrobial compounds as combined hurdles. J. Dairy Sci. 91.

55. Vardanyan, R.S., Hruby, V.J., 2006. Antifungal drugs. Available at https://doi.org/10.1016/B978-044452166-8/50035-2.

56. Ahmad, V., Khan, M.S., Jamal, M.S., Alzohairy, M.A., Karaawi, A.A., Siddiqui, M.U., 2017. Antimicrobial potential of bacteriocins: in Therapy, Agriculture and Food Preservation. Int J Antimicrob Agents. 49(1): 1–11. doi: 10.1016/j.ijantimicag.2016.08.016.

57. Williams, G.C., Delves-Broughton J., 2003. Nisin, pp. 4128–4135. In Caballero B (ed.), Encyclopedia of Food Sciences and Nutrition, 2nd edition.

58. Stevens, M., Vollenweider, S., Lacroix, C., Zurich, E.T.H., 2011. The potential of reuterin produced by Lactobacillus reuteri as a broad spectrum preservative in food. Available at https://doi.org/10.1533/9780857090522.1.129.

Chapter 7

Chemistry of Natural Food Additives and Preservatives

Tahira Batool Qaisrani,[a] Muther Mansoor Qaisrani,[b] Muhammad Shahbaz,[c] and Khurshied Ahmad Khan[d]

[a]*Faculty of Agricultural Sciences, Ghazi University, Dera Ghazi Khan, Punjab, Pakistan*

[b]*Khawja Fareed University of Engineering and Information Technology, Rahim Yar Khan, Punjab, Pakistan*

[c]*Muhammad Nawaz Sharif University of Agriculture, Multan, Punjab, Pakistan*

[d]*National University of Science and Technology, Islamabad, Pakistan*

tqaisrani@gudgk.edu.pk

Natural, healthy, and safe food is the requirement of all eras. Natural food additives add value without causing health hazards. The consumer demand for green foods is building new space for the use of natural food preservatives and additives at the industrial level. Natural food additives and preservatives are the gifts of nature while chemical additives are under discussion due to their safety impact on human health. Increasing health awareness and concerns of the consumer worldwide is crafting the threat of chemical additives. Additives from both sides, natural or chemical/synthetic sources,

Natural Food Preservatives

Edited by Sourish Bhattacharya and Heba Hassan Abd-El Azim Salama

ISBN 978-981-4968-64-5 (Hardcover), 978-1-003-36776-5 (eBook)

www.jennystanford.com

exhibit their effect on the food traits. Hence, an insight into the chemistry of a food additive and preservative; the domineering parameter for effective use in foods is the key requisite. The main sources of natural preservatives are plants (fruits, vegetables, spices, herbs, seeds, and others) and microorganisms, mainly the lactic acid bacteria (LAB). Plant-based sources provide phytoalexins, organic acids, phenolic compounds, and essential oils. From the microbial side, LAB is providing nisin, pediocin, sakacin, lactacin, acidophilucin, plantaricin, plantacin, helveticin, lactocin, and others. In this chapter, the chemistry of these compounds will be discussed for their effective and safe use in food products.

7.1 Introduction

Dynamic human nature is always in search of a new variety of foods. A food processor is in continuous search for inventing or modifying new methods/techniques to fulfill human requirements and eagerness. But in any circumstances, there is no compromise on the safety of food. Natural foodstuff with additional ingredients from natural sources is being preferred after the successful era of chemically synthesized food additives. The common perception is that natural resources are certainly potent to provide healthy and safe food. Briefly, the food and food materials from natural resources are considered safe from the emergence of food processing, centuries ago. Different substances are in use to increase the food value with respect to appearance, flavor, softness, and also for nutritional enrichment in the foods. For this purpose, both natural and synthetic food components are added to foods to obtain certain specific traits [1, 2].

Natural food additives (flavoring, acidulants, antioxidants, antimicrobial, etc.) are preferred for value addition in conventional and industrially processed foods. The key aspect is a presumption of the community that these are not hazardous to health but rather support the human social life and monetary affairs. Reduction in risk factors (that are associated with chemical/synthetic substances in food) stands responsible for decreasing the diseases and disorders, thereby reducing the expenditures against combating the ailments [1–3].

The consumer demand for green foods is creating space for natural food additives and preservatives in small and large industries side by side. Natural resources for the purpose are the blessings of nature especially when synthetic additives are facing debate because of their possible and reported negative impact on safety parameters. Increasing health awareness among consumers (all inclusive) is creating a barrier to further use of chemical additives. The food additives from either natural resources or synthesized artificially certainly affect the identical characteristics of food. The quality and safety traits of foods better gratify consumer preferences. A better understanding of the chemical properties, structure, and reactions of food additives and preservatives is mandatory as a domineering parameter for its effective and safe use in foods [1, 3].

Plant origin is capable to provide phytoalexins, organic acids, phenolic compounds, essential oils, and many others [4–7]. From the microbial side, lactic acid bacteria are providing a variety of antimicrobial chemicals. Being produced through natural resources, these are potent to be registered in natural food additives and preservatives. Primarily known as bacteriocins, these compounds include nisin, pediocin, sakacin, lactacin, acidophilucin, plantaricin, plantacin, helveticin, lactocin, and a few more [8–11]. In this chapter, the chemistry of these compounds, sources, and uses as food additives and preservatives will be discussed for their safe practice in food products.

7.2 Natural Additives/Preservatives from Plant Sources

7.2.1 Organic Acids

Organic acids from natural sources or as synthetic chemicals are widely added to foods to perform multitasking functions in food processing. Extracted from natural foods (Table 7.1), these are traditionally being used in foods mainly as acidulants and to improve the flavor in both deserts and spicy food recipes [12, 14]. As the food spoilage and food pathogenic microorganisms cannot grow at lower pH conditions, the organic acids are also exploited as food preservatives [7]. Their functionality may be as a sequestrant, buffer to adjust the acidity, and antioxidants. Ascorbic acid, lactic acid, acetic acid, malic acid, and citric acid are frequently used organic acids [7, 12, 15].

Table 7.1 Organic acids and their food sources

Name of organic acid	Food source
Lactic acid	Milk and butter digestion, food fermentation
Acetic acid	Vinegar
Citric acid	Citrus fruits: lemon, orange
Malic acid	Apple
Tartaric acid	Grapes, pineapples
Oxalic acid	Tea, pepper
Tannic acid	Tea
Caffeotannic acid	Coffee
Benzoic acid	Cranberries, plums
Butyric acid	Decomposition of butter

Fruits such as berries, apples, grapes, plums, and from the citrus family (lemon, orange, etc.) are the principal source of food-oriented acids. Furthermore, they voluntarily regulate the human system being the natural food component compared to some synthetic organic acids that the human metabolic system cannot unsurprisingly recognize, thus leading to enragement and soreness. The synthetic compounds require strict legislation. Owing to increased industrial demand for natural/green additives, the primary metabolites and partaking low molecular weight organic acids (lactic, acetic, and citric acids) are generated through fermentation. Biotechnologically built organic acids are attracting attention due to legislature and growing awareness for eco-friendly chemicals [12, 14].

7.2.1.1 Lactic acid/milk acid

Lactic acid as obtained from milk is also called milk acid. Its IUPAC name is 2-hydroxypropanoic acid. The chemical formula of lactic acid is $C_3H_6O_3$ and it commonly exists in the form of α-hydroxy carboxylic acid (Fig. 7.1). It is chiral and hence its two enantiomeric forms are L- and D-lactic acid, where L-lactic acid mainly elaborates in human metabolism. It is white in a solid state (Fig. 7.2) but forms a colorless solution when dissolved (yogurt). It is tenfold more acidic than acetic acid because of hydrogen bonding between the α-hydroxyl and the carboxylate group (making it alpha-hydroxy acid). The molar mass of lactic acid is 90.078 g/mol; it is hygroscopic and soluble in water. Its melting and boiling points are 18°C and 122°C, respectively [2, 14, 20].

Predominantly, in industry lactic acid is produced through fermentation (70–90%) performed by lactic acid bacteria which convert simple carbohydrates such as glucose, galactose, and sucrose to lactic acid. It is also produced chemically from acetaldehyde. In 2006, the global production of lactic acid reached 275,000 tons with an average annual growth of 10%. Lactate is the conjugate base of lactic acid.

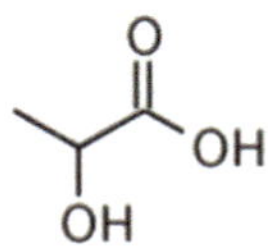

Figure 7.1 Molecular structure of lactic acid.

Figure 7.2 Isolated, crystalline structure of lactic acid.

It has been given E number E270 in the class of food preservatives. Hence, lactic acid is a pronounced food additive serving as pH manager (to give the effect of food preservation), as a flavoring compound (enhances and protects flavor of dairy, meat, and vegetable products), and as umami flavorings (enhances a broad range of savory flavors). Its use through direct acidification in some foods (pickles, yogurt, cultured milk products, and LAB cultured meat products) is unique [8, 20] and when added as an additive in others (drinks, salad dressings, meat preservation, etc.), it is similar to other umami flavoring organic acids. The salts of lactic acid, that is, sodium or potassium lactate, function as antimicrobial agents to control foodborne pathogens and increase water retention capacity in meat and fish [7, 12, 14].

7.2.1.2 Citric acid

Structurally, it is solid and crystalline (Fig. 7.3). The molecular formula is 2-hydroxypropane-1,2,3-tricarboxylic acid (Fig. 7.4), and the chemical formula is $C_6H_8O_7$. A high temperature above 175°C causes decarboxylation. Although this acid is considered a weak acid, it is stronger than representative carboxylic acids. It is obtained as monohydrate with 210.14 g/mol molar mass and as anhydrous with 192.123 g/mol of molar mass. The flash point of this acid is 155°C. Citric acid is potent to form a variety of mono-, di-, tri-, and mixed esters because of the availability of its three carboxylic acid groups [24]. Its solubility in water at 20°C is 59.2% w/w basis. It is also soluble in alcohol, ether, ethyl acetate, and acetone but insoluble in toluene, benzene, and trichloromethane [12, 18].

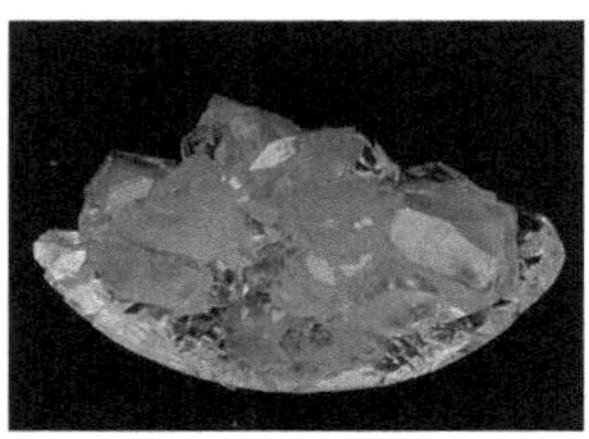

Figure 7.3 Isolated crystalline citric acid.

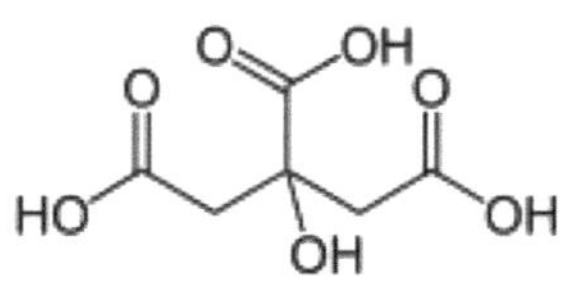

Figure 7.4 Molecular structure of citric acid.

It is recognized as an antioxidant with E330 number. Citric acid is well recognized as green/natural preservative in the food industry and is also used to enhance the antimicrobial efficacy of the preservatives. It provides a representative sour taste to foods, especially being the choice for juices, thirst-quenching drinks, and carbonated beverages. Its properties such as buffering, solubility, and flavoring make it a component of food additives for syrup preparations [7, 12, 21].

7.2.1.3 Tartaric acid/winestone

Tartaric acid is in white crystalline form (Fig. 7.5) and is dicarboxylic aldric acid. The IUPAC name preferred for tartaric acid is 2,3-dihydroxybutanedioic acid with chemical formula $C_4H_6O_6$ and structural formula (Fig. 7.6) as $HO_2CCH(OH)CH(OH)CO_2H$. Its molar mass is 150.087 g/mol with 1.79 g/mL (H_2O) of density; the melting point varies from 171°C to 174°C. Some different forms of tartaric acid are levo-tartaric acid, dextro-tartaric acid, meso-tartaric acid, and racemic acid. The properties vary with its various forms. Different forms of this acid exhibit difference in their solubility in water. For L- or D-tartaric, DL-racemic, and meso, it is 1.33, 0.21, and 1.25 kg/L, respectively [7, 12].

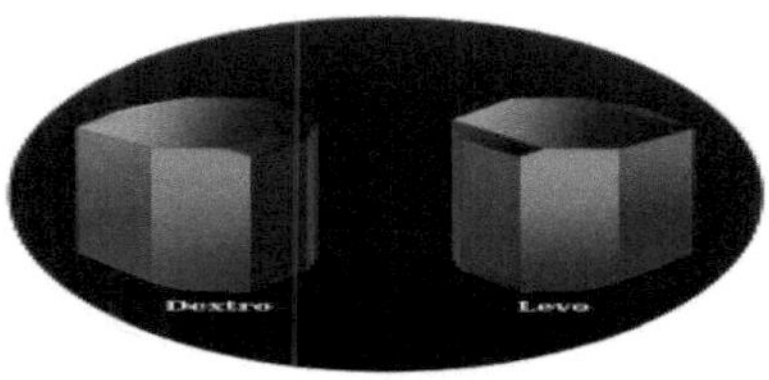

Figure 7.5 Image of D- and L-tartaric acid crystals through an optical microscope.

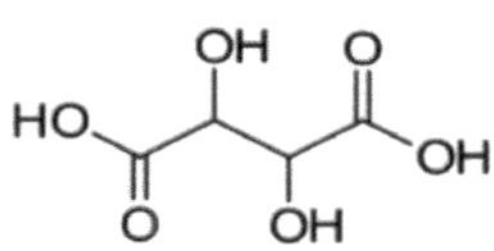

Figure 7.6 Molecular structure of tartaric acid.

Natural sources of tartaric acid are many plants, mainly grapes, pineapple, and tamarind. Interestingly, higher levels of tartaric acid have been found in citrus fruits produced through organic farming compared to that of the conventional forming system [14]. Its title roles are as acidulant (wine), antioxidant (beverages as healthy acid), buffering agent (chemically, being a dihydroxy derivative of succinic acid), and also leavening agent (if added in combination with baking soda) [4, 7, 22].

7.2.1.4 Malic acid

The chemistry of malic acid reveals it with IUPAC name of 2-hydroxybutanedioic acid. It is an organic substance with the chemical formula $C_4H_6O_5$ existing in L- and D-enantiomer forms (Figs. 7.7 and 7.8) but the naturally occurring form is only the L-isomer. The other names for malic acid are hydroxybutanedioic acid, 2-hydroxysuccinic acid, L/D-malic acid, and ±-malic acid. It is colorless acid with 134.09 g/mol of molar mass. Its solubility in water is 558 g/L at 20°C with a density value of 1.609 g/cm^3. Malate is the term used for ester and the salt of malic acid. Malic acid is responsible for the acid content of a variety of fruits such as sour apples, apricots, pears, plums, cherries and berries, quince, and even in citrus fruits. It has attained E number E296 in the class of food preservatives. The melting point of this organic acid is 130°C showing different melting properties from other acids being used in the food. It is melted readily and therefore conformed into molten hard candies requiring no additional water, thereby resulting in the increased shelf life of the product [7, 12, 23].

Figure 7.7 D- and L-malic acid.

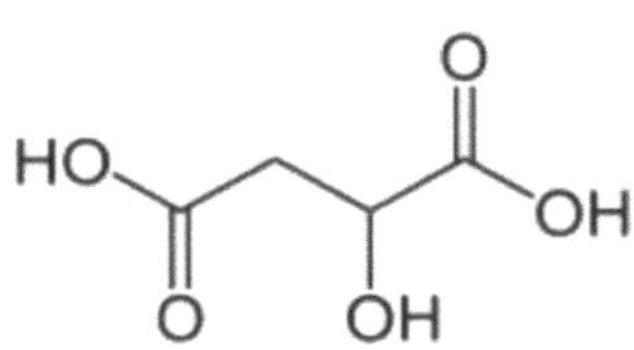

Figure 7.8 Molecular structure of malic acid.

Its amount required to get the effect of acidic taste is very less than that of citric acid. Therefore, it is merely used in low calorie drinks. But its sour taste has an extended effect and persists longer marking an additional composed taste. The addition of malic acid in syrups, deserts, and water ices improves the pH control, acid sugar ratio, and fruity flavor of the commodities [7, 12].

7.2.1.5 Acetic acid

Other names of acetic acid are methane carboxylic acid, hydrogen acetate, and vinegar (diluted form). The vinegar contains acetic acid not less than 4% by volume. The systematic name given is ethanoic acid. Its chemical formula is CH_3COOH (Fig. 7.9), which also can be written as $C_2H_4O_2$. The molar mass of acetic acid is 60.052 g/mol with 1.049 g/cm^3 and 1.27 g/cm^3 in liquid and solid forms (Fig. 7.10), respectively. The melting and boiling points are 16–17°C and 118–119°C, respectively, with a flash point of 40°C. Liquid acetic acid is a polar solvent like water; therefore, it dissolves inorganic salts and sugars. It also can dissolve nonpolar compounds like oils as it is miscible with polar and nonpolar solvents. An acetate is an ionic form obtained by the loss of H^+ from its acid, and thus it is referred to as a salt or an ester of acetic acid [26]. The pH of domestic vinegar is 2.4. It appears colorless in liquid form with an intense vinegar odor. It is used in food as an acidulant and condiment with food additive code number E260 [7, 12].

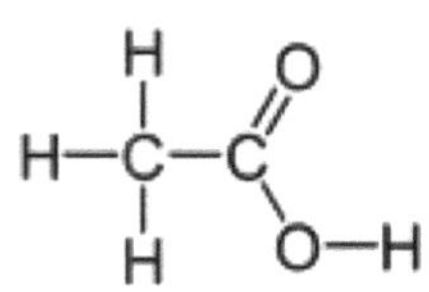

Figure 7.9 Molecular structure of acetic acid.

Figure 7.10 Isolated crystalline acetic acid.

Vinegar, which is the dilute acetic acid, is usually formed by fermentation (carried out by acetic acid bacteria) following the oxidation of ethanol. The bacteria from the genus *Acetobacter* and *Clostridium acetobutylicum* are used for this purpose. Naturally, acetic acid is produced as a result of the spoilage of fruits and other foods. The biological production of acetic acid financially accounts for 10% of world produce that principally is utilized for the production of vinegar. The food purity laws necessitate that vinegar used as a food additive should be of biological origin [12, 25].

7.2.2 Phytoalexins

Phytoalexins are synthesized in various plant species in response to attacking organisms, hence working as toxins for invaders to that plant. This term was, therefore, used by plant pathologists. They perform antimicrobial (in the form of toxins) and antioxidant (controlling free radicals) functions by quickly accumulating at infection zones. Chemically, they are secondary metabolites and possess diverse traits depending on plant species and therefore are inhibitors of a wide spectrum. Generally, they are naturally produced chemicals by plant sources and are included in the classes of alkaloids, phenols, terpenoids, and glycol-steroids. These can be obtained naturally for their food additive/preservative role by induction of the specific related nematodes, fungi, viruses, and bacteria [27, 28, 34] to respective plant parts. The chemistry and functional properties along with plant sources of the phytoalexins given in Table 7.2 will be helpful in further research studies to reveal their specific role as natural food additives and preservatives.

7.2.2.1 Pisatin

Pisatin is formed in the harvested pea pods in retaliation to the attack of many fungi (*Monilinia fructicola*, *Ascochyta pisi*, etc.) or in response to pod injury. Further higher levels of pisatin are produced in ambient temperatures under aerobic conditions. Pisatin is a phenolic ether and very mild phytoalexin. It holds a chromocoumarin ring system with the molecular formula $C_{17}H_{14}O_6$ and its molecular weight is 314 g/mol [12, 27, 34].

Table 7.2 Names and structures of phytoalexins [12, 27–37]

Name	Structure	Name	Structure
Pisatin	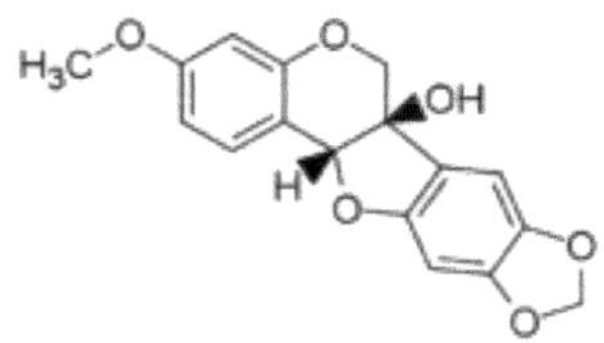	Cicerin	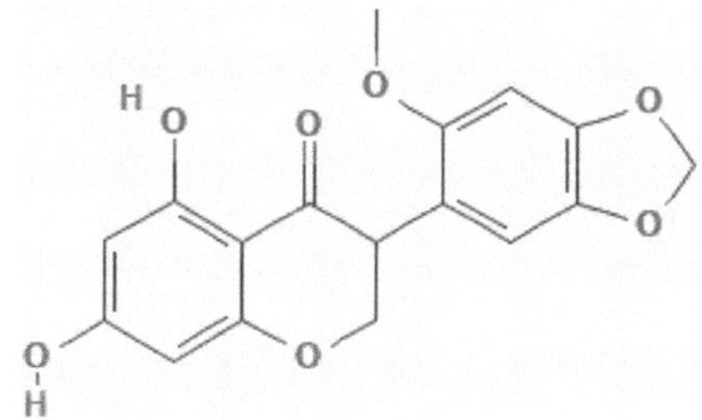
Gossypol	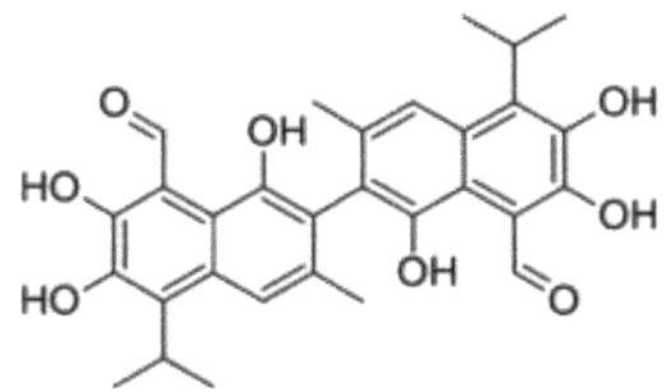	Medicarpin	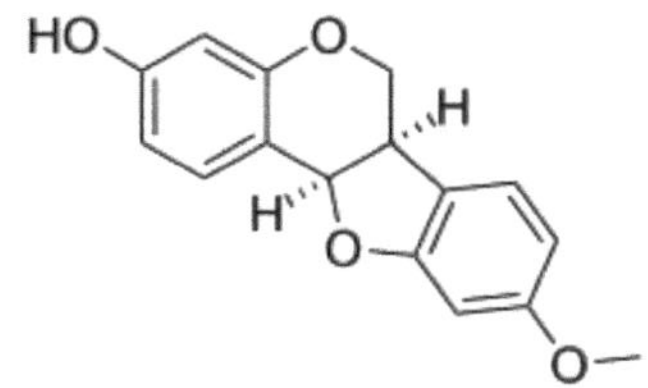
Isocoumarin	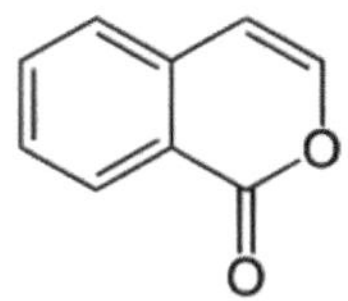	Orchinol	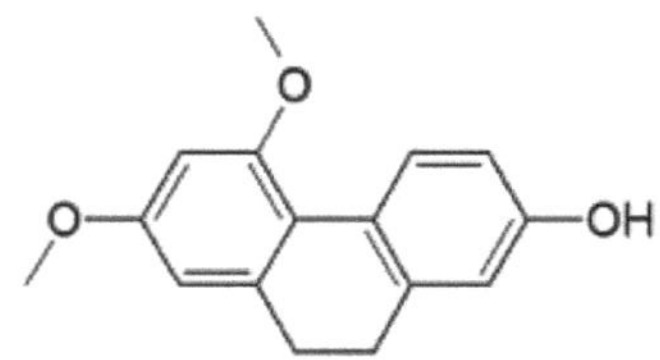

(*Continued*)

Table 7.2 (*Continued*)

Name	Structure	Name	Structure
Rishitin	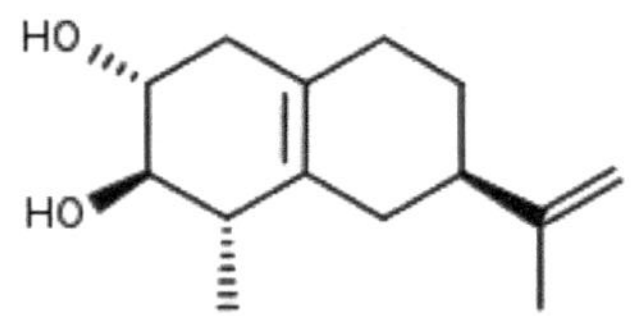	Phaseollin	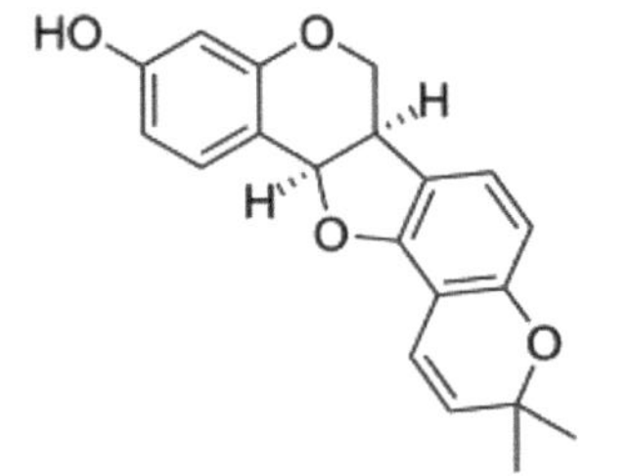
Ipomeamarone	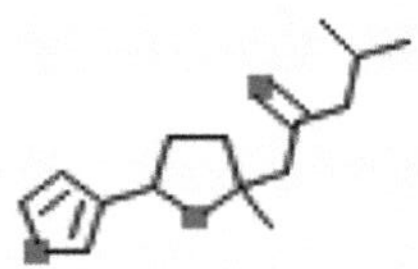	Capsidiol	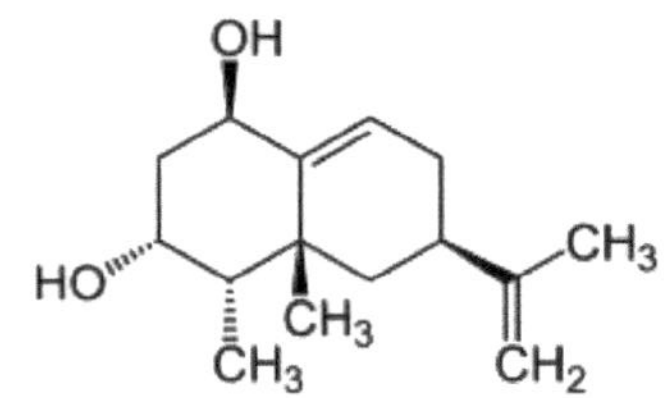
Glyceollin I	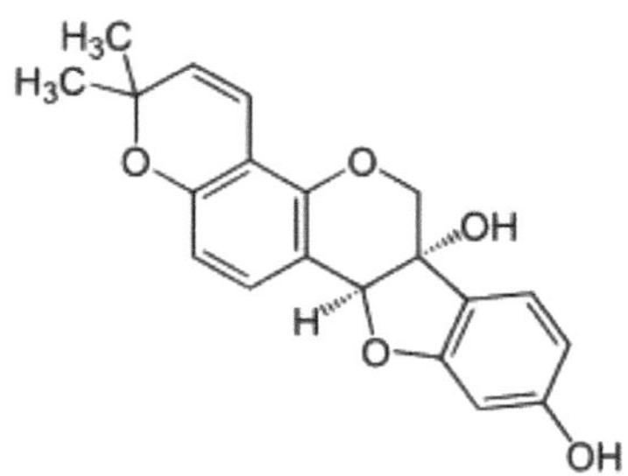	Allexin	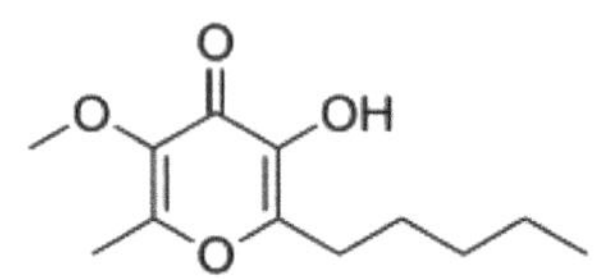

7.2.2.2 Cicerin

Cicerin like pisatin is composed of phenolic compounds. Its source is *Cicer* (gram) when attacked by the fungus *Ascochyta rabiei* causing disease Ascochyta blight. Cicerin disseminates into immunization droplets in pod tissues of a gram [29].

7.2.2.3 Medicarpin

Medicarpin, another antifungal compound, is produced in the leaves of alfalfa, groundnut, and legumes and provides defense against the fungi (*Colletotrichum phomoides, Puccinia arachidis, Helminthosporium turcicum*, etc.). The identity of this antifungal (fungi may be pathogenic and non-pathogenic) compound has been marked as (–) dimethyl-homopterocarpin. Isocoumarin is synthesized in carrot root tissues contaminated with non-pathogen species (*Cyratocystis fimbriata*) and pathogens (*Fusarium oxysporum, Rhizopus stolonifer*, etc.) [1, 30]

7.2.2.4 Orchinol

The IUPAC nomenclature of orchinol is 5,7-dimethoxy-9,10-dihydrophenanthrene-2-ol with the chemical formula $C_{16}H_{16}O_3$ and a molar mass of 256.29 g/mol. Orchinol is a phenanthrene isolated from *Orchis militaris* and Loroglossumhircinum if infected with fungi *Rhizoctonia repens*. It has a broad antimicrobial range and is difficult to specify [1, 12, 31].

7.2.2.5 Rishitin

Rishitin has the IUPAC name (1S,2R,3R,7R)-1-methyl-7-(prop-1-en-2-yl)-1,2,3,4,5,6,7,8-octahydronaphthalene-2,3-diol with the molecular formula given as $C_{14}H_{22}O_2$. It is a non-sesquiterpene alcohol with 222.328 g/mol molar mass. This substance is compiled by tubers of *Solatium tuberosum* var. *rishiri* infected with spores of *Phytophthora* and by potato tubers infected with *F. solani* f. sp. *phaseoli* [12, 32].

7.2.2.6 Phaseollin

Phaseollin has been given the IUPAC name 3,3-dimethyl-6b,12b-dihydro-3H,7H-furo[3,2-c:5,4-f']dichromen-10-ol with molecular formula $C_{20}H_{18}O_4$ and the molar mass value is 322.360 g/mol [30].

Phaseollin is comparable to pisatin in chemical and functional aspects. It was found in harvested bean pods bearing the non-disease-causing fungus *M. fructicola* and in bean leaves contaminated with the bacterium *Pseudomonas phaseolicola*. Recently, similar composites phaseollidin, phaseollinisoflavan, and kievitone were identified from diffusates containing various types of microbiota such as bacteria, fungi, and viruses [27, 33].

7.2.2.7 Ipomeamarone

Chemistry of ipomeamarone indicates that it is (2r-*cis*)-isomer, a molecule of the class of heteroaromatic compounds. Its IUPAC name is 1-[5-(furan-3-yl)-2-methyloxolan-2-yl]-4-methylpentan-2-one with molecular formula $C_{15}H_{22}O_3$ and molecular weight is 250.338 g/mol. Practically, it is insoluble in water and having only 0.041 g/L solubility value and possessing very low acidic activity. Ipomeamarone, a tetrahydrofuran, is synthesized in the roots of sweet potato (*Ipomoea batata*) infected with the fungus *Ceratocystis funbriata* causing black rot disease. Biosynthesis of ipomeamarone may be induced by *Helicobasidium mompa*, the violet root fungus, poisonous chemicals, and pectinase enzyme. It functions as an inhibitor of electron transport and energy transfer reactions and thus prevents mycelial growth, sporulation, and protein synthesis of the pathogen and in presence of toxic chemicals. Nevertheless, ipomeamarone is noxious to both non-pathogenic and pathogenic fungi and to some bacteria [12, 35].

7.2.2.8 Capsidiol

The phytoalexin capsidiol is amongst the sesquiterpenoids and a bicyclic terpene. The IUPAC name of this compound is 4,4a-dimethyl-6-(prop-1-en-2-yl)-1,2,3,4,4a,5,6,7-octahydronaphthalene-1,3-diol and the chemical formula is $C_{15}H_{24}O_2$. It is rarely soluble in water (0.85 g/L) and has an average molecular weight of 236.3499 g/mol. This natural fungicide is produced in response to infection with fungi in sweet pepper fruits, tobacco, and green vegetables [27, 34].

7.2.2.9 Glyceollin I

Glyceollin I is chemically isopentyl mercaptan from the class of alkylthiols exhibiting nearly neutral pH activity.

The IUPAC name is 17,17-dimethyl-3,12,18-trioxapentacyclo[$11.8.0.0^{2},0^{10}.0^{4},0^{9}.0^{14},0^{19}$]henicosa-1(13),4(9),5,7,14(19),15,20-heptaene-6,10-diol and it has the chemical formula $C_{20}H_{18}O_5$. It is practically insoluble in water (0.06 g/L) and its molecular weight is 338.3539 g/mol. It has been discovered in soybean and alcoholic beverages. It tastes like onion and glue. The production of glyceollin I occurs in response to infection against soybean cyst nematode (*Heterodera glycines*) and the fungus (*P. megasperma* f. sp. *glycinea*), only in the resistant varieties [12, 36].

7.2.2.10 Allixin

Allixin is named 3-hydroxy-5-methoxy-6-methyl-2-pentyl-4H-pyran-4-one in the IUPAC system of nomenclature. Its molecular formula is $C_{12}H_{18}O_4$ and molecular weight averages 226.272 g/mol. It is a non-sulfur-containing molecule composed of a γ-pyrone skeleton structure. It was isolated from garlic. Allixins can be used for other health-related benefits as allixins contains antioxidant and antimicrobial properties in food processing sector and can be used as a nutrient additive in functional foods [12, 37].

7.2.3 Polyphenols

Polyphenols are a class of natural organic compounds abundantly found in plants. They are attributed to the presence of several phenol units (many hydroxyl groups on aromatic rings) making these varied in structure. The polyphenols are moderately water-soluble compounds possessing a molecular weight of 500–4000 Da. Chemical traits refer to them as antioxidants [12]. Hence, they can be added to foods (meat, fish, etc.) and drinks (fruit-based drinks and beer) to foil the oxidation of lipids and other susceptible compounds to improve shelf steadiness. Polyphenols include flavonoids, tannic acid, curcumin, ellagic acid, ellagitannin, tellimagrandin II, theaflavin-3-gallate, and so on [4, 12, 38]. A detailed description is given in Table 7.3.

Table 7.3 The source, chemistry, and structure of some polyphenols [4, 41–49]

Name	IUPAC name	Molecular formula	Molar mass (g/mol)	Solubility in water (g/L)	Plant source	Function in food	Structure
Tannic acid	1,2,3,4,6-penta-*O*-{3,4-dihydroxy-5-[(3,4,5-trihydroxybenzoyl)oxy]benzoyl}-D-glucopyranose	$C_{76}H_{52}O_{46}$	1701.19	2850	Tara pods, Gallnuts Aleppo oak	Food ingredient but not as additive	(*See* Figure 1)
Curcumin	(1*E*,6*E*)-1,7-bis(4-hydroxy-3-methoxyphenyl)hepta-1,6-diene-3,5-dione	$C_{21}H_{20}O_6$	368.385	0.0057	Turmeric	Dietary supplement, flavouring agent, and as colorant with E100	(*See* Figure 2)
Ellagic acid	2,3,7,8-Tetrahydroxy[1]benzopyrano[5,4,3-*cde*][1]benzopyran-5,10-dione	$C_{14}H_6O_8$	302.197	0.82	Berries, walnuts, pecans, pomegranate	Nutriceutical; anticancer, antioxidation	(*See* Figure 3)
	(10aR,11S,12aR,25aR,25bS)-23-{[(10aR,11S,12aR,25aR,25bS)-2,3,4,5,6,7,17,18,19,20,21,22-Dodecahydroxy-9,15,24,27-tetraoxo-9,10a,11,12a,13,15,24,25a,25b-,27-decahydrodibenzo[g,i]dibenzo[6',7':8',9'][1,4]						

Ellagitannin	dioxecino[2',3':4,5]pyrano[3,2-b][1,5]dioxacycloundecin-11-yl]oxy}-2,3,4,5,6,7,17,18,19,20,21,22-dodecahydroxy-9,15,24,27-tetraoxo-9,10a,11,12a,13,15,24,25a,25b,27-decahydrodibenzo[g,i]dibenzo[6',7':8',9'][1,4]dioxecino[2',3':4,5]pyrano[3,2-b][1,5]dioxacycloundecin-11-yl3-({(10aR,11S,12aR,25aR,25bS)-2,3,4,5,6,7,17,18,19,20,21,22-dodecahydroxy-9,15,24,27-tetraoxo-11-[(3,4,5-trihydroxybenzoyl)oxy]-9,10a,11,12a,13,15,24,25a,25b,27-decahydrodibenzo[g,i]dibenzo[6',7':8',9'][1,4]dioxecino[2',3':4,5]pyrano[3,2-b][1,5]dioxacycloundecin-23-yl}oxy)-4,5-dihydroxybenzoate	$C_{116}H_{76}O_{74}$	2653.79	2.79	Raspberry, cloves, guava, pomegranate	Antioxidant	(*See* Figure 4)

(*Continued*)

Table 7.3 (*Continued*)

Name	IUPAC name	Molecular formula	Molar mass (g/mol)	Solubility in water (g/L)	Plant source	Function in food	Structure
Tellimagrandin II	(11a*R*,13*S*,14*R*,15*S*,15a*R*)-2,3,4,5,6,7-Hexahydroxy-9,17-dioxo-9,11,11a,13,14,15,15a,17-octahydrodibenzo[*g*,*i*]pyrano[3,2-*b*][1,5]dioxacycloundecine-13,14,15-triyl tris(2,3,4-trihydroxybenzoate)	$C_{41}H_{30}O_{26}$	938.66	1.19	Asian herb bennet, clove	Anti-viral and antioxidant	(*See* Figure 5)
Theaflavin-3-gallate	(2*R*,3*R*)-5,7-Dihydroxy-2-{3,4,6-trihydroxy-5-oxo-8-[(2*R*,3*R*)-3,5,7-trihydroxy-3,4-dihydro-2*H*-1-benzopyran-2-yl]-5*H*-benzo[7]annulen-2-yl}-3,4-dihydro-2*H*-1-benzopyran-3-yl 3,4,5-trihydroxybenzoate	$C_{36}H_{28}O_{16}$	716.604	0.12	Black tea, green tea	Antioxidant as food ingredient	(*See* Figure 6)

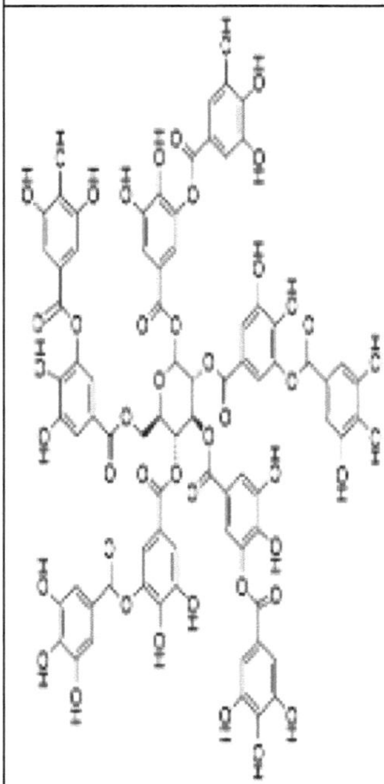

Figure 1

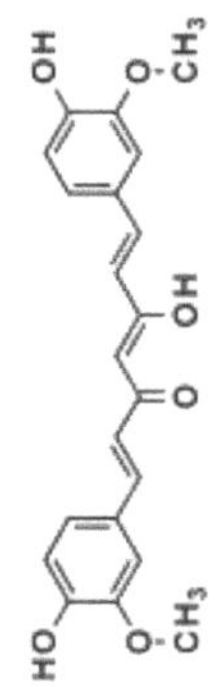

Figure 2

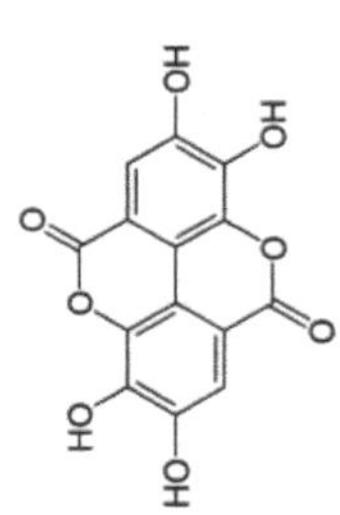

Figure 3

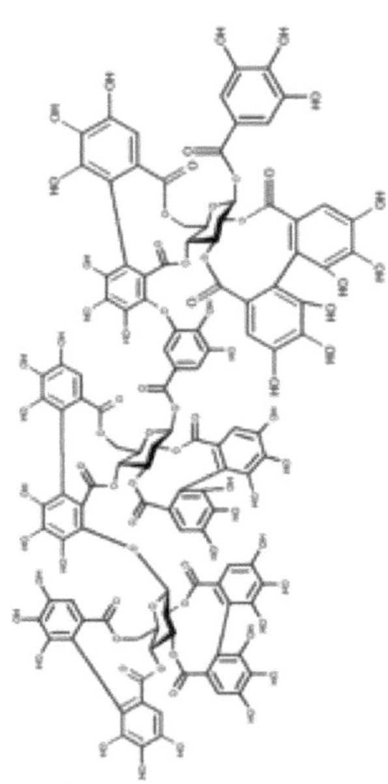

Figure 4

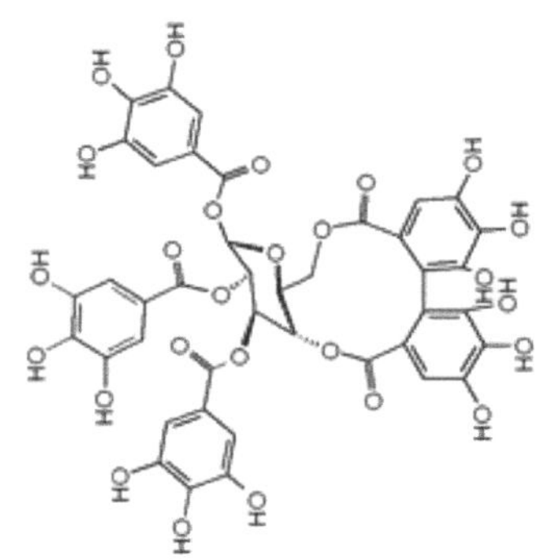

Figure 5

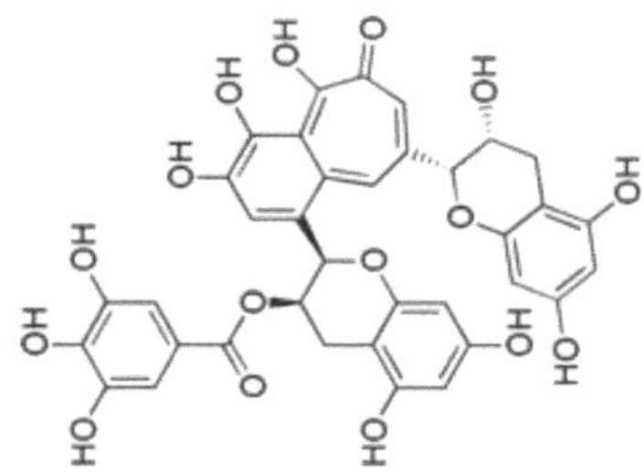

Figure 6

Flavonoids particularly abundant in foods are cyanidin (red fruits and berries), proanthocyanidins (grapes, apples, cocoa), catechins (tea – green and black, fresh fruits), daidzein (soybean), quercetin (onion, tea, apples), and hesperetin (citrus fruits) [2]. Phenolic acids include caffeic acid. Plant polyphenols may serve as promising natural food antioxidant and antimicrobial agents compared with sulfiting agents to keep the quality of fish and fish-based products [41]. The multi-direction use of natural polyphenols makes them attractive for their safe use. They can be antioxidants, appealing agents (carotenoids and flavonoids), structural materials (lignin), signaling substances (salicylic acid), and protecting compounds (tannins and phytoalexins) [4, 37, 39].

The phenolic compounds can be added to foods to extend their shelf life (preservative), to turn them into functional food (nutrient), or to incorporate them into food packaging. These can be extracted from natural sources, such as olives, grapes, fruits, vegetables, rice, spices, herbs, tea, and algae, among others. The phenolics have the capacity of retarding the microbial invasion in some products and avoiding the putrefaction of others, mainly fruits and vegetables [38–40]. Therefore, in different food products, phenolic compounds have started to replace chemical additives and are perceived to be safer and hence claimed to alleviate safety concerns.

There are many methods of polyphenols application, but the main is direct addition. Spraying, coating, and dipping treatment of food are currently applied to the product prior to packaging as valid options and can be used to delay or inhibit the oxidation and growth of microorganisms. However, in food applications, the phenolic compounds could be influenced by food components, processing, and storage [40].

In vitro antioxidant activity of many fruits and vegetables can be observed. The oregano oil is a rich source of (as more than 50% of this oil) phenolic compounds (primarily carvacrol and thymol). Cinnamon extract is generally recognized as safe as a flavoring component in foods providing an anti-bacterial effect but also inhibits mold growth and decreases the production of aflatoxin up to 99% [38, 39]. Analysis techniques for polyphenols include paper chromatography, thin layer chromatography, reversed-phase liquid chromatography, high performance liquid chromatography, fractionation, spectroscopy, or chemical assay. Extraction can be achieved through solvents (water, methanol, methanol/formic acid, etc.). Black tea yields polyphenolic content up to 20% of its fresh weight, while for most of fruits (grapes, berries, etc.), this content is from 0.2% to 0.3% of fresh weight [4, 38, 40, 41].

7.2.4 Essential Oils

Commercially 300 different essential oils obtained from nearly 2000 plant species are in use, but a very few of them are used as food additives. Chemical substances data provides 3000 essential oils on the list. The term "essential" is used with the logic of their belonging to the plant of which they were a part like garlic oil. Hence, they should not be confused with the term used for essential nutrients [45].

Chemically the essential oils (EOs) are either terpenes or phenylpropanoids (Fig. 7.11). From various classes of terpenes (hemiterpenes, monoterpenes, sesquiterpenes, and diterpenes), mostly the EOs are composite blends of sesquiterpenes ($C_{15}H_{24}$) and monoterpenes ($C_{10}H_{16}$). EOs contain phenylpropanoids generally in the form of phenols and phenol ethers where the side chain may be shortened (C1) in some cases. The principal phenylpropanoid representatives are safrole, anethole, and eugenol possessing double bonds in the side chain. Physical traits offer essential oils usually as eloquent and portable liquids, and rarely these are solid (orris) or semisolid (guaiac wood) at ambient temperature. The plant origin, species, and the part of the plant where stored are responsible for the identical aroma of the essential oils. They are found in different parts of plants such as in the glands, oil ducts, trichomes, and resin ducts [45–47].

The extraction of EOs is carried out through various techniques including extraction (organic solvent extraction, ultrasonic extraction, supercritical carbon dioxide extraction, solvent-free microwave extraction, high-pressure solvent extraction,), distillation (hydro-distillation, microwave-assisted distillation, steam distillation), the phytonic process, expression, microwave hydrodiffusion, and gravity [46].

Essential oils are composed of low molecular weight compounds (less than 500 Da) with lower relative density than that of water. They display optimal rotation and higher refractive index because of many asymmetrical compounds and also may form resinous molecules through polymerization after being oxidized. Generally, they are considered hydrophobic and hence easily soluble in organic solvents and miscible with fats and oils. These are present in different parts of plants such as flowers, seeds, wood, root, bark twig, and leaf. Chemically, these are concentrated hydrophobic, liquid, volatile, and semi-volatile organic chemicals evaporating at normal temperatures [45, 46].

Table 7.4 Essential oil and its use in foods [17, 45–49]

Essential oils	Food additive function	Essential oils	Food additive function
Balsam of Peru	Flavoring compound	Turmeric	Antimicrobial flavoring compound
Coriander	Flavoring compound	Coffee	Flavoring compound
Ginger	Antimicrobial, antioxidant	Caraway	Flavoring compound
Calamus	Flavoring compound	Curry	Flavoring compound
Galangal	Flavoring compound	Oregano	Antimicrobial
Cannabis	Flavoring compound	Orange	Flavoring compound
Rose oil	Supplement and flavoring compound	Garlic	Antimicrobial, antioxidant
Coconut	Culinary purpose	Asafoetida	Flavoring compound

They are also referred to as ethereal oils and aetheroleum. Though they are termed oils, essential oils are not firm oils. They are lipophilic and immiscible to water, thereby possessing poor solubility in water. They are obtained from plant parts through techniques such as distillation, cold pressing, expression, florasol extraction, and solvent extraction, resin tapping, wax embedding, and cold pressing. Organic solvents such as pure ethanol and polyethylene glycol can be used for their dilution. The essential oils have been used as flavoring, food supplements, and preservatives in traditionally opted techniques for decades [45].

They as part of natural sources are considered generally recognized as safe (GRAS) by the Food and Drug Administration. Essential oils have identical odors representing their plant source and hence used as flavoring agents or aroma enhancers. Recent research workers are converging on their antimicrobial antioxidant and food preservation aspects [12, 47, 48]. Most commonly used essential oils in foods, beverages, and confectioneries are given in Table 7.4.

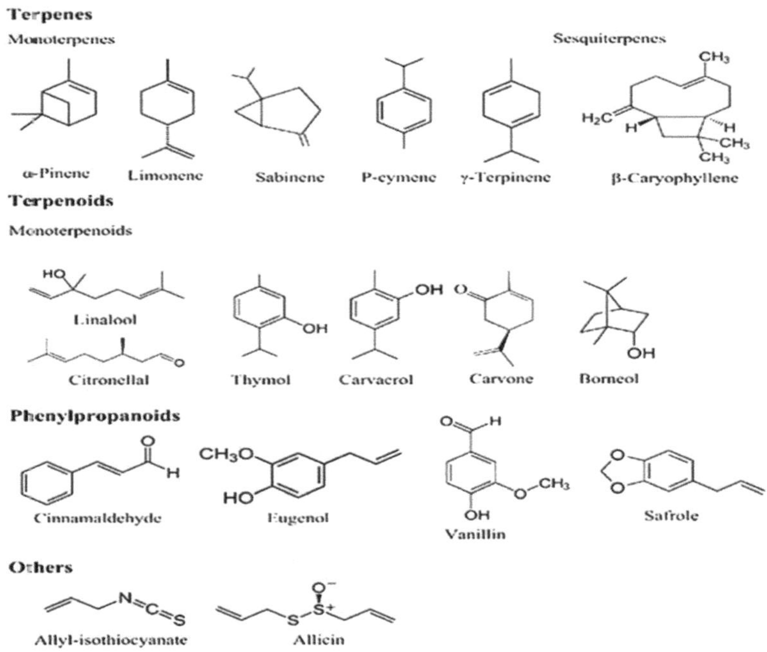

Figure 7.11 Terpenes and phenylpropanoids [16].

7.3 Natural Additives/Preservatives from Lactic Acid Bacteria

Fermentation of various foods (milk, meat, fruits, and vegetables) is carried out by culturing these with lactic acid bacteria (LAB). This practice is continuing in domestic (traditionally) and industrial zones equally. LAB works as protective culture against food poisoning and food spoilage microorganisms in foods such as milk and milk products, meat, fish, seafoods, fruits, and vegetable products. The LAB when cultured as selective fermentation produce certain chemicals for preventing other bacterial invasions in that food. The compounds produced in this way are termed "bacteriocins" [1, 8].

The method of application varies from direct culturing of bacteriocin-producing LAB as a part of the fermentation process, in purified/crude form, and as adjuvant culture in milk and milk products for successful control of pathogens in dairy products. The bacteriocins produced by LAB are of various types and research is open to their chemical, physical, and bactericidal aspects as natural and safe food additives and preservatives. As produced by a natural source, these are generally regarded as safe by the FDA for their use in food processing [1, 8–10].

The examples of few bacteriocins from LAB are nisin, pediocin, sakacin, lactacin, acidophilucin, plantaricin, plantacin, helveticin, lactocin, and so on [8, 50–56]. Although a large number of such compounds have been recognized, their chemical traits and application in foods for specific purposes are still under consideration.

7.3.1 Bacteriocins

7.3.1.1 Nisin

Nisin is produced by lactic acid bacteria the bacterium *Lactococcus lactis*. It was first isolated in the 1930s and has been produced from natural substances since the 1950s but was approved as a food additive in the USA in the 1960s. At industrial levels, it is biologically obtained through fermentation of natural substances (milk lactose or dextrose), not synthesized chemically [8].

Its chemical formula is $C_{143}H_{230}N_{42}O_{37}S_7$ and boiling point is 2966°C. It is powdery material with a molar mass of 3354.07 g/mol and a density of 1.402 g/mL at 25°C. It was included in the class of generally recognized as safe compounds by the FDA in 2016 [44]. Amongst the bacteriocins, the use of nisin as a food preservative (E number E234) is trending in numerous countries recently. Nisin actively functions as a preservative in amounts measured nearly to parts per billion levels (level range ~1–25 ppm) using techniques such as chromatography and diffusion bioassays [8, 12]. It is a broad spectrum bacteriocin operative against various Gram-positive bacteria, including lactic acid bacteria, *Bacillus cereus*, *C. botulinum*, *Listeria monocytogenes*, *Staphylococcus aureus*, and so on. Particularly, it works against spores. Gram-negative bacteria are protected by their outer membrane but may become susceptible to nisin action after a heat shock or when this is coupled with some chelating agent like EDTA [12, 50, 51].

7.3.1.2 Sakacin

Sakacins are formed by the bacteria *Lactobacillus sakei* and usually grouped with other LAB produced natural bacteriocins. Most of the sakacins have been discovered in natural food backgrounds (sakacin B from the dry cured sausage). Sakacins as bacteriocins produced in foods are of many types including sakacin A, B, G, K, M, P, Q, R, T, and Z. Some are also named with numbers like Sakacin 670. These are organic compounds overall belonging to the class of polypeptides with a varying amino acid profile. Only Sakacin A and Sakacin P are discussed here as the remaining are still lacking sufficient published data to present [1, 8, 10].

7.3.1.2.1 *Sakacin A*

It is among the organic compounds and typically from the class called aspartic acid and its derivatives. The chemical formula for this molecule is $C_{12}H_{24}N_4O_4$ with IUPAC name as 3-amino-3-[(1-{[tert-butyl(methyl)carbamoyl]amino}ethyl)carbamoyl]propanoic acid. Its average molecular weight is reported as 288.3434 g/mol and the solubility rate in water is 2.45 g/L [12, 52].

7.3.1.2.2 *Sakacin P*

Sakacin P is categorized as fatty amides (carboxylic acid amide derivatives of fatty acids) from the general class of organic compounds. Thereby, it is grouped in fatty amide lipid substances. It exhibits higher hydrophobic properties and hence practically insoluble in water with a value of 4.04 g/L and fairly neutral. It has the molecular formula of $C_9H_{16}N_2O_4$ and the IUPAC name is *tert*-butyl *N*-(1-carbamoyl-1-oxopropan-2-yl)carbamate with an average molecular weight of 216.2343 g/mol. Sakacin P is a ribosomally produced, small-sized, heat-stable polypeptide.

Research studies for the production of sakacins on biotechnology grounds in food media are in line. The practical use of sakacins in modern food processing and food chemistry is growing against *Listeria* (in sausages), cured meat-based foods (ham and cold cuts), cheeses, and also covering other LAB fermented foodstuffs. They can control the undesirable growth of certain bacteria causing malodor, ropiness (dairy products), and sliminess [12, 52, 53].

7.3.1.3 Plantaricin BN

This class of bacteriocin belongs to organic compounds identified as tetracarboxylic acids and their derivatives. These carboxylic acids precisely possess four carboxyl groups. The plantaricin BN is a very weak, basic, nearly neutral compound. The IUPAC name given to plantaricin BN is 3-[1,4,4-tris(2-carboxyethyl)-2,3-dioxocyclododecyl]propanoic acid with chemical formula as $C_{24}H_{36}O_{10}$. The molecular weight reported is 484.5366 g/mol and the molecule shows a water solubility value of 0.13 g/L, that is, very poor [1, 8, 12].

7.3.1.4 Pediocin AcH

Pediocin AcH is an organic compound from a class of polypeptides containing 10 or more amino acid residues. Pediocin AcH may be neutral. Structurally, it is an aromatic hetero-polycyclic compound with the molecular formula $C_{196}H_{297}N_{61}O_{60}S_5$ and the average molecular weight is 4628.2 g/mol [54].

7.3.1.5 Halocins

Halocin bacteriocins are biologically synthesized by halophilic Archaea and hence are like archaeocin. The different types of halocins have been discovered after their initial product was coined in 1982. Chemically, these may be protein (large halocins) or polypeptides (small halocins). The biological role of halocins as a natural antimicrobial agent is under consideration for controlling food spoilage during industrial food processing [44, 55, 56].

7.4 Conclusion

Green foods are trending with the concept of healthy foods. Minimally processed food is another term in line. The emphasis on the consumer is the health aspect. The processor is in need of compounds that can serve the consumer at the minimum cost difference in processing. The objectives of both concerned corners can be satisfied with the use of components extracted from natural food sources like plants and LAB. These can be directly used as functional food components or as added food components. The chemistry of a few components are provided in this chapter. Research corner provides very little information on such components for their proven use as a food additive. Therefore, dedicated work is required from the agencies such as research, legislation processing, and also involving the nutritionists to develop safe foods using natural substances, thereby replacing the chemically synthesized food additives.

References

1. Baines, D. and Seal, R. (eds.). (2012). Natural Food Additives, Ingredients and Flavourings. Woodhead Publishing Series in Food Science, Technology and Nutrition. https://doi.org/10.1533/9780857095725.1
2. Bearth, A., Cousin, M., and Siegrist, M. (2014). The consumer's perception of artificial food additives: Influences on acceptance, risk and benefit perceptions. Food Quality and Preference, 38, 14–23. http://dx.doi.org/10.1016/j.foodqual.2014.05.008
3. Bruna, G. O. L, Thais, A. C. C., and Ligia A. C. C. (2018). Food additives and their health effects: A review on preservatives sodium benzoate. African Journal of Biotechnology, 17(10), 306–310. https://doi.org/10.5897/AJB2017.16321
4. Lin, D., Xiao, M., Zhao, J., Li, Z., Xing, B., Li, X., Kong, M., Li, L., Zhang, Q., Liu, Y., Chen, H., Qin, W., Wu, H., and Chen, S. (2016). An overview of plant phenolic compounds and their importance in human nutrition and management of type 2 diabetes. Molecules, 21, 1374. https://doi.org/10.3390/molecules21101374
5. Shand, R. F. and Leyva, K. J. (2008). Archaeal antimicrobials: an undiscovered country. Archaea: New Models for Prokaryotic Biology. Caister Academic Press. ISBN: 978-1-904455-27-1

6. Braïek, O. B. and Smaoui, S. (2021). Chemistry, safety, and challenges of the use of organic acids and their derivative salts in meat preservation. Journal of Food Quality, 2021, 1–20. https://doi.org/10.1155/2021/6653190
7. Moghaddam, M. and Mehdizadeh, L. (2017). Chemistry of essential oils and factors influencing their constituents. In: Soft Chemistry and Food Fermentation. Elsevier Inc. http://dx.doi.org/10.1016/B978-0-12-811412-4.00013-8
8. Bacteriocin. https://en.wikipedia.org/wiki/Bacteriocin
9. Bouayed, J. (ed.). (2012). Nutrition, Well-Being and Health. InTech. ISBN: 978-953-51-0125-3
10. https://www.fda.gov/food/food-ingredients-packaging/overview-food-ingredients-additives, 2010.
11. Quitmann, H., Fan, R., and Czermak, P. (2014). Acidic organic compounds in beverage, food, and feed production. Advances in Biochemical Engineering/Biotechnology, 143, 91–141. https://doi.org/10.1007/10_2013_262
12. International Union of Pure and Applied Chemistry. (2014). Nomenclature of Organic Chemistry: IUPAC Recommendations and Preferred Names 2013. The Royal Society of Chemistry. ISBN: 978-0-85404-182-4. doi: 10.1039/9781849733069
13. Valgimigli, L. (ed.). (2012). Essential Oils as Natural Food Additives: Composition, Applications, Antioxidant and Antimicrobial properties, pp. 1–466. ISBN: 978-1-62100-241-3
14. Duarte, A. M., Caixeirinho, D., Miguel, M. G., Sustelo, V., Nunes, C., Fernandes, M. M., and Marreiros, A. (2012). Organic acids concentration in citrus juice from conventional versus organic farming. Acta Horticulturae, 933, 601–606. doi:10.17660/actahortic.2012.933.78
15. Ukeh, D. A. and Mordue, A. J. (2009). Plant based repellents for the control of stored product insect pests. Biopesticides International, 5, 1–23.
16. Shukla, P., Akshay, S., and Ashok, S. (2017). Food additives from an organic chemistry perspective. J. Bio. Org. Chem., 1, 70–79. doi: 10.15406/mojboc.2017.01.00015
17. Pandey, A. K., Kumar, P., Singh, P., Tripathi, N. N., and Bajpai, V. K. (2017). Essential oils: Sources of antimicrobials and food preservatives. Frontiers in Microbiology, 7, 2161. doi: 10.3389/fmicb.2016.02161

18. https://www.who.int/news-room/fact-sheets/detail/food-additives, 2018.

19. Endres, H. (2009). Technische Biopolymere, p. 103. München: Hanser-Verlag. ISBN: 978-3-446-41683-3

20. Groot, W., van Krieken, J., Slekersl, O., and de Vos, S. (2010). Chemistry and production of lactic acid, lactide and poly (lactic acid). In Auras, R., Lim, L., Selke, S. E., and Tsuji, H. (eds.). Poly (Lactic acid). Hoboken: Wiley. ISBN: 978-0-470-29366-9

21. https://foodb.ca/unearth/q?utf8=%E2%9C%93&query=Citric+acid&searcher=compounds&button

22. https://foodb.ca/unearth/q?utf8=%E2%9C%93&query=Tartaric+acid&searcher=compounds&button=

23. https://foodb.ca/unearth/q?utf8=%E2%9C%93&query=Malic+acid&searcher=compounds&button=

24. Bergeron, R. J., Xin, M. G., Smith, R. E., Wollenweber, M., McManis, J. S., Ludin, C., and Abboud, K. A. (1997). Total synthesis of rhizoferrin, an iron chelator. Tetrahedron, 53, 427–434. doi:10.1016/S0040-4020(96)01061-7

25. https://foodb.ca/unearth/q?utf8=%E2%9C%93&query=Acetic+acid&searcher=compounds&button=

26. Hendrickson, J. B., Cram, D. J. and Hammond, G. S. (1970). Organic Chemistry. 3rd ed. p. 128, London: McGraw-Hill, Inc.

27. https://en.wikipedia.org/wiki/Phytoalexin

28. https://foodb.ca/compounds/FDB011855

29. Ye, X. Y., Ng, T. B., and Rao, P. F. (2002). Cicerin and arietin, novel chickpea peptides with different antifungal potencies. Peptides, 23(5), 817–822. https://doi.org/10.1016/s0196-9781(02)00005-0

30. Yang, X., Zhao, Y., Hsieh, M., Xin, G., Wu, R., Hsu, P., Horng, L., Sung, H., Cheng, H., and Le. K. (2017). Total synthesis of (+)-medicarpin. Journal of Natural Products, 80, 3284–3288.

31. Arditti, J., Flick, B. H., Ehmann, A., and Fisch, M. H. (1975). Orchid phytoalexins. II. Isolation and characterization of possible sterol companions. American Journal of Botany, 62, 738–742.

32. Glazener, J. A. and Wouters, C. H. (1981). Detection of rishitin in tomato fruits after infection with Botrytis cinerea. Physiological Plant Pathology, 19, 243–248. https://doi.org/10.1016/S0048-4059(81)80027-6

33. https://foodb.ca/unearth/q?utf8=%E2%9C%93&query=Phaseollin&searcher=compounds&button=
34. Huang, J. S. (2001). Accumulation of Phytoalexins as a Resistance Mechanism. In: Plant Pathogenesis and Resistance. Springer, Dordrecht. https://doi.org/10.1007/978-94-017-2687-0_9
35. Yoshinosuke Usuki, Y., Deguchi, T., and Iio H. (2014). A new concise synthesis of (+)-ipomeamarone, (−)-ngaione, and their stereoisomers. Chemistry Letters, 43,1882–1884. 10.1246/cl.140811
36. Boué, S. M., Isakova, I. A., Burow, M. E., Cao, H., Bhatnagar, D., Sarver, J. G., Shinde, K. V., Erhardt, P. W., and Heiman, M. L. (2012). Glyceollins, soy isoflavone phytoalexins, improve oral glucose disposal by stimulating glucose uptake. Journal of Agricultural and Food Chemistry, 60, 6376–6382. https://doi.org/10.1021/jf301057d
37. Moriguchi, T., Matsuura, H., Itakura, Y., Katsuki, H., Saito, H., and Nishiyama, N. (1997). Allixin, a phytoalexin produced by garlic, and its analogues as novel exogenous substances with neurotrophic activity. Life Sciences, 61(14), 1413–1420. https://doi.org/10.1016/s0024-3205(97)00687-5
38. Cory, H., Passarelli, S., Szeto, J., Tamez, M., and Mattei, J. (2018). The role of polyphenols in human health and food systems: A mini-review. Frontiers in Nutrition, 5, 87. https://doi.org/10.3389/fnut.2018.00087
39. Song, N., Ma, L., Wang, W., Sun, H., Lei Wang, L., Baldwin, I. T., and Wu, I. 2021. Capsidiol, a defensive sesquiterpene produced by wild tobacco in response to attack from the fungal pathogen Alternaria alternata, is regulated by an ERF2-like transcription factor. bioRxiv, 573675. doi: https://doi.org/10.1101/573675
40. Sara, M., Javier, R. P., Manuel, C. B., Daniel, M. V., and Jonathan, D. (2017). Application of phenolic compounds for food preservation: food additive and active packaging. In: Phenolic Compounds - Biological Activity. ISBN: 978-953-51-2959-2
41. Maqsood, S., Benjakul, S., and Shahidi, F. (2013). Emerging role of phenolic compounds as natural food additives in fish and fish products. Critical Reviews in Food Science and Nutrition, 53, 162–179. https://doi.org/10.1080/10408398.2010.518775
42. Ziogas V., Tanou G., Molassiotis A., Diamantidis G., and Vasilakakis M. (2009). Antioxidant and free radical-scavenging activities of phenolic extracts of olive fruits. Food Chemistry, 120, 1097–1103. doi: 10.1016/j.foodchem.2009.11.058

43. Romero, C., Brenes, M., García, P, and Garrido A. (2002). Hydroxytyrosol 4-β-d-glucoside, an important phenolic compound in olive fruits and derived products. Journal of Agricultural and Food Chemistry, 50, 3835–3839. doi: 10.1021/jf011485t

44. Fahim, H. A., Khairalla, A. S., and El-Gendy, A. O. (2016). Nanotechnology: A valuable strategy to improve bacteriocin formulations. Frontiers in Microbiology, 7, 1385. doi:10.3389/fmicb.2016.01385

45. Djilani, A. and Dicko, A. (2012). The therapeutic benefits of essential oils. In: Nutrition, Well-Being and Health, pp. 155–178. doi: 10.5772/1864

46. Nedorostova, L., Kloucek, P., Kokoska, L., Stolcova, M., and Pulkrabek, J. (2009). Antimicrobial properties of selected essential oils in vapour phase against food borne bacteria. Food Control, 20, 157–160. doi: 10.1016/.2008.03.007

47. Mimica-Dukic, N., Bozin, B., Sokovic, M., Mihajlovic, B., and Matavulj, M. (2003). Antimicrobial and antioxidant activities of three Mentha species essential oils. Planta Medica, 69, 413–419. doi: 10.1055/s-2003-39704

48. Preedy, V. (ed). (2015). Essential Oils in Food Preservation, Flavor and Safety. Elsevier Scientific Publications. Academic Press. doi: https://doi.org/10.1016/C2012-0-06581-7

49. Hyldgaard, M., Mygind, T., and Meyer, R. L. (2012). Essentials oils in food preservation: Mode of action, synergies and interactions with food matrix components. Frontiers in Microbiology, 3, 12. doi: 10.3389/fmicb.2012.00012

50. Adem, G., Oulahal, N., Catherine Joly, C., and Degraeve, P. (2016). Nisin as a food preservative: Part 1: Physicochemical properties, antimicrobial activity, and main uses. Critical Reviews in Food Science and Nutrition, 56(8), 1262–1274. doi: 10.1080/10408398.2013.763765

51. Gharsallaoui, A., Joly, C., Oulahal, N., and Degraeve P. (2016). Nisin as a food preservative: Part 2: Antimicrobial polymer materials containing nisin. Critical Reviews in Food Science and Nutrition, 56, 1275–1289. doi: 10.1080/10408398.2013.763766

52. Trinetta, V., Morleo, A., Sessa, F., Iametti, S., Bonomi, F., and Ferranti, P. 2012. Purified sakacin A shows a dual mechanism of action against *Listeria* spp: Proton motive force dissipation and cell wall breakdown. FEMS Microbiology Letters, 334, 143–149. https://doi.org/10.1111/j.1574-6968.2012.02630.x

53. Mathiesen, G., Huehne, K., Kroeckel, L., Axelsson, L., and Eijsink, V. G. (2005). Characterization of a new bacteriocin operon in sakacin P-producing Lactobacillus sakei, showing strong translational coupling between the bacteriocin and immunity genes. Applied and Environmental Microbiology, 71(7), 3565–3574. https://doi.org/10.1128/AEM.71.7.3565-3574.2005
54. https://foodb.ca/unearth/q?utf8=%E2%9C%93&query=pediocin&searcher=compounds&button=
55. Rodríguez-Valera, F., Juez, G., and Kushner, D. J. (1982). Halocins: Salt-dependent bacteriocins produced by extremely halophilic rod. Canadian Journal of Microbiology, 28, 151–154. doi: 10.1139/m82-019
56. Birbir, M., Eryilmaz, S., and Ogan, A. (2004). Prevention of halophilic microbial damage on brine cured hides by extremely halophilic halocin producer strains. Journal of the Society of Leather Technologists and Chemists, 88, 99–104.

Chapter 8

Naturally Occurring Antimicrobial System

Kawthar Mahmoud Aseel,[a] Rehab Omer Elnour,[b] and Sabah Mohamed Abbas Hamza[b]

[a]*Sudan University of Science and Technology, College of Engineering and Technology of Industries, Department of Chemical Engineering, Food Processing Engineering, Sudan*

[b]*King Khalid University, KSA, Faculty of Science and Arts, Dhahran Aljanoub, Department of Biology, Abha-KSA*

relgezouly@kku.edu.sa

There is a need for natural antimicrobial compounds which may replace synthetic preservatives. Characteristic antimicrobials from various sources are utilized to save food from waste and pathogenic microorganisms. Plants (spices and flavors, leafy foods, seeds, and leaves) are the primary wellspring of antimicrobials and contain numerous fundamental oils that have a safeguarding impact against various microorganisms. For the most part, spices and flavors

Natural Food Preservatives
Edited by Sourish Bhattacharya and Heba Hassan Abd-El Azim Salama

ISBN 978-981-4968-64-5 (Hardcover), 978-1-003-36776-5 (eBook)
www.jennystanford.com

contain numerous fundamental oils and the models incorporate rosemary, sage, basil, oregano, thyme, cardamom, and clove. These fundamental oils are successful against numerous pathogenic and waste microorganisms such as *Salmonella*, *Escherichia coli*, *Listeria monocytogenes*, *Campylobacter* spp., and *Staphylococcus aureus* and help to expand their quality and rack solidness. These antimicrobial mixtures are likewise utilized in mix with consumable food coatings and hinder the capacity of microorganisms to develop on the outside of food and food items.

8.1 Introduction

Today, food handling is everyone's anxiety and it is elusive for any individual who has not experienced an undesirable snapshot of foodborne sickness at any rate once in the previous year. As per the report of WHO in 2005, there were about 1.8 million passing brought about by loose bowels (foodborne sickness), and these illnesses were because of the utilization of tainted food and water [1]. The fundamental driver of foodborne sicknesses is the utilization of food polluted by microbial microorganisms, poisons, or radioactive parts. At the point when certain microorganisms or microbes sully food, they can cause foodborne ailment or at times called "food contamination." Foodborne ailments are mellow; however, at times they can even be destructive [2].

Foodborne microorganisms (*Clostridium botulinum*, *S. aureus*, *C. jejuni*, *Bacillus cereus*, *L. monocytogenes*, *Cryptosporidium*, *E. coli* O157:H7, etc.) are the primary concern in regard to the well-being of food [3]. Food can contain microbiological microorganisms that cause diseases or inebriations, or substance specialists that cause intense or persistent inebriations. With extraordinary reference to meat and meat items, *Salmonella*, *E. coli*, *L. monocytogenes*, and *Campylobacter* are the primary pathogenic life forms [4, 5].

There is an increment in the utilization of new food with the buyer interest for the prepared to-eat food and the longing to lead a sound way of life. The difficulties related to the utilization of new food are short stockpiling life and its relationship with foodborne illnesses.

To stay away from these difficulties, there is a business pressing factor of utilizing substance additives that forestall the development of food deterioration specialists; however, the expansion in the utilization of these compound additives is adversely seen by the buyer [4].

8.2 Reasons to Use Natural Antimicrobial Compounds

In numerous nations around the world, there is a quickly developing interest for climate cordial, safe additives to be utilized for gentle food protection. Customary food protection procedures affect the allure of new food items, and counterfeit additives are progressively being prohibited [6].

As an outcome, an assortment of new or negligibly prepared, exceptionally short-lived vegetables has arisen in the market having gone through milder safeguarding procedures, for example, a mix of refrigeration and changed climate bundling.

Gentle safeguarding methods can handle item deterioration brought about by microorganisms somewhat, chiefly on the grounds that they are utilized in adherence with the "obstacle innovation" (consolidated cycles) idea [7]. However, it is currently turning out to be more obvious that potential well-being dangers may happen with the gentle protection frameworks because of the endurance and development of certain foodborne microbes. Of uncommon concern are cold-lenient (psychrotrophic) microbes, similar to *L. monocytogenes*, *Yersinia enterocolitica*, and *Aeromonas hydrophila*, which may develop to levels of worry during the long time frame of realistic usability of these short-lived food sources.

Mesophilic microorganisms (i.e., *Salmonella* spp., *S. aureus*, enteropathogenic *E. coli*, and *B. cereus*) represent a well-being danger when temperature misuse happens. Hence, there is a critical requirement for the presentation of extra well-being factors with these gentle protection strategies [6].

For quite a while, compound additives, for example, sorbate and benzoate, have been utilized as dependable additive variables to control various microbial perils. Notwithstanding, such mixtures do not fulfill the idea of "regular" and "sound" food that purchasers like and that the food business, thus, necessities to make. The negative response to synthetic additives in our general public is emphatically expanding, in spite of the way that such mixtures are at this point key in food preparation. Accordingly, the substitution of synthetics by more regular options must be applicable when fundamental (i.e., when the compound options are not at this point satisfactory) and conceivable (i.e., when common substitutes are in reality [eco-] toxicologically protected to utilize and compelling by and by). The need is underlined by numerous individuals in the agro-industry, law-making bodies, and customer associations. The chance is upheld by numerous investigations performed by scholastics and food industrialists. Plainly characteristic choices are not generally as intense as existing synthetics and the smart utilization of consolidated cycles might be essential for ideal usefulness. Likewise, it is obvious that even normal options should pass legislatory examination and that the mark "common" ought not be mistaken for characteristic well-being [6].

Nature is notable to contain various sorts of antimicrobial mixtures that assume a significant part in the common guard or rivalry frameworks of a wide range of living beings, going from microorganisms to creepy crawlies, creatures, and plants. This mixture is considered the most achievable substitute for compound food additives considering reasonable, legislatory, and moral angles [6].

As to the improvement of regular antimicrobial mixtures from plants (aggregately called green synthetic compounds) for food conservation, research is currently centered around the possible utilization of phytoalexins, natural acids, and phenols. What's more, encouraging outcomes have been achieved with fundamental oils from spices and fragrant plants. Such fundamental oils comprise combinations of esters, aldehydes, ketones, and terpenes with a wide range of antimicrobial action. The toxicological premise of numerous spices and flavors just as their dynamic segments has been contemplated and frequently they are known to be food grade or even GRAS (by and large perceived as protected) [6].

Over the most recent couple of many years, consideration of new common antimicrobial mixtures has emerged because of an adjustment of shopper inclinations and the increment in the quantity of safe microorganisms. All the more as of late, utilitarian items and particularly common useful fixings have delighted in a lift in buyer interest. These items are typically liked by the customer over engineered fixings—a pattern that is becoming in the food business as well as in different areas [8].

During the most recent many years, two principal patterns have animated interest in new characteristic antimicrobial builds. In the first place, common fixings with additive properties have encountered an expanding request, in substitution of the utilization of manufactured fixings, to forestall microbial defilement as they are more secure, eco-friendly, and have a wide spectrum of activities and they keep away from a portion of the results related with engineered antimicrobials [9, 10]. Second, in late many years, an increment in the quantity of microorganisms (microscopic organisms and parasites) impervious to antimicrobial medications has happened. This issue is viewed as now as a general medical issue since conventional anti-toxins and antifungals have lost viability [11].

These days, it has been exhibited that manufactured antimicrobials, similar to sodium benzoate, sodium nitrite, or sorbic acid, utilized in the food business, can deliver adverse consequences on the health of the consumers. Consequently, the quest for new antimicrobials obtained from regular sources is acquiring significance [12].

Antimicrobials could be incorporated for both the food items as novel useful fixings and in the bundling material [13]. In the main case, antimicrobials could be added to the food straightforwardly to forestall the decay because of the development of microorganisms [12].

8.3 Natural Antimicrobials of Plant Origin

8.3.1 Herbs and Spices

Humans have long used herbs and spices for various reasons such as food additives, flavors, and preservatives. These are considered the most commonly used natural antimicrobials against various pathogens. The antimicrobial activity of the herbs and spices relies upon the sort of fundamental oil present in them, the food type in which they must be utilized, and the kind of microorganism [14–16].

A portion of the antimicrobial mixtures that are available in herbs and spices are eugenol, thymol, thymol and carvacrol, vanillin, allicin, cinnamic aldehyde, and allyl isothiocyanate that are, individually, present in cloves, thyme, oregano, vanilla, garlic, cinnamon, and mustard [17].

The efficacy of fundamental oils from spices and flavors relies on their synthetic construction, specifically to the presence of hydrophilic useful gatherings, for example, hydroxyl bunches [18].

Fundamental oils from clove, oregano, rosemary, thyme, sage, and vanillin are the best containing the phenolic bunches [19]. They have inhibitory movement against Gram-positive than Gram-negative microorganisms.

Essential oils have high fume pressure and can arrive at pathogenic microorganisms through gas or fluid stages. Numerous examinations have demonstrated the antimicrobial effectiveness of essential oils against a few pathogenic and deterioration microflorae. In any case, the productivity of fundamental oils relies on the pH, stockpiling temperature, and convergence of oxygen.

8.3.2 Phytoalexins

Phytoalexins are characterized as host-combined, low-subatomic weight, wide range antimicrobial mixtures whose amalgamation from removed antecedents is prompted in plants in light of microbial disease or treatment of plant tissues with a scope of normally happening or engineered, fake mixtures (biotic or abiotic elicitors) [20]. At this point, more than 200 distinctive phytoalexins have been recognized in excess of 20 plant families. Phytoalexins are wide range anti-infection agents, for the most part dynamic against phytopathogenic parasites. As opposed to the preformed prohibitins, illness obstruction due to phytoalexins is a unique interaction, requiring a new combination of optional metabolites. Likewise, the compounds dependable for the union of phytoalexins are themselves blended because of openness to organisms or other successful upgrades [6]. Elicitors, the mixtures setting off the amalgamation of phytoalexins, range in nature from bacterial proteins [21] to parasitic unsaturated fats to have plant-determined oligosaccharides [3]. Besides their antimicrobial action, the phytoalexins are often formulated as anti-growth, despite the fact that action has additionally been accounted for toward microorganisms. Gram-positive microorganisms have been discovered to be more touchy than Gram-negative microbes. Isoflavonoids, which are depicted by the basic structure of the C_6 - C_3 - C_6, are among the main classes of phytoalexins and concentrates on their application outside the common sources have been attempted [22]. The leguminosa are known for the creation of isoflavonoid phytoalexins, for instance, pisatin (Fig. 8.1 right) from *Pisum sativum*, phaseollin from *Phaseolus vulgaris*, and glyceollin from *Glycine max*. It has attracted a lot of attention in the creation of isoflavonoid phytoalexins in plant cell and tissue culture [23] and this could be a technique for large-scale counterfeit creation when a plant itself has no solid business potential.

For basically related mixtures in the gathering of isoflavonoid phytoalexins, it was tracked down that an increment in lipophilicity associates decidedly with expanded antifungal movement [24]. Terpenoid phytoalexins, for example, rishitin (Fig. 8.1 left), are predominantly found in the group of Solanaceae, for example, in potato tubers.

It was found that the activity of rishitin in the laboratory against bacteria was inhibited due to the low levels of the divalent Ca^{2+} and Mg^{2+} cations, indicating that these compounds act on the cytoplasmic membrane of the target microorganisms.

Another significant gathering of phytoalexins, likewise alluded to as infection or pathogenesis-related proteins, involves chitinases, thionins, zeamatins, and thaumatins [25]. A portion of these proteins are associated with the blend of other phytoalexins or phenolic compounds as constitutive or inducible catalysts. Others supposedly have an immediate antimicrobial impact. Since they are proteins, they would be totally processed in the stomach and hence would not affect the strength of a customer. Chitinases target chitin, a significant segment of the cell mass of most phytopathogenic parasites, and furthermore of the skeletal construction of most spineless creatures, for example, creepy crawlies and bugs. Solid plants typically contain low degrees of chitinase, yet their creation is actuated after a microbe assault. The actuated chitinase aggregates either intracellularly or in the intercellular space, where their action is required. Since vertebrates and higher plants do not contain chitin, no unfriendly effect is known, supporting the allure of chitinases for fungus control [6].

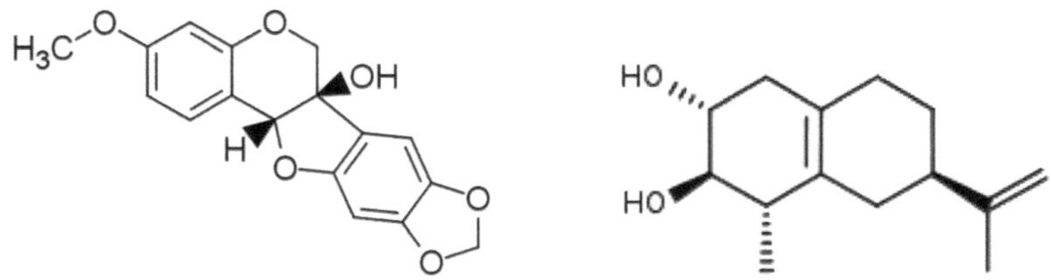

Figure 8.1 Phytoalexins from *P. sativum* (right) and *Solanum tuberosum* (left).

8.3.3 Organic Acids

Citric, succinic, malic, and tartaric acids are normally found in organic products (e.g., citrus, rhubarb, grapes, and pineapples) and vegetables (e.g., broccoli and carrots). Through their utilization as acidulants or cancer prevention agents in food varieties, their antimicrobial properties give an extra advantage.

The antimicrobial action of the different acids is broadly archived [7]. They target cell dividers, cell films, metabolic compounds, protein blend frameworks, and hereditary material. In this way, they are dynamic against a wide scope of microorganisms. The natural acids contained in harvests may well add to the common yield opposition. Numerous natural acids or their subordinates are now applied as food additives.

8.3.4 Phenolic Compounds

Phenolic compounds are characterized as mixtures containing hydroxylated fragrant rings. The phenolic compounds are classified into three groups: simple phenols and phenolic acids (e.g., *p*-cresol, 3-ethylphenol, hydroquinone, protocatechuic, vanillic, gallic, syringic, and ellagic acids), hydroxycinnamic acid derivatives (e.g., *p*-coumaric, caffeic, ferulic, and sinapic acids), and flavonoids [26]. The latter group is the most important single group of phenolics in food, comprising catechins, proanthocyanins, anthocyanidins, flavones, flavonols, and their glycosides. Finally, tannins, a polymeric form of phenolics, are an important group of plant phenolics, unified by the common ability to precipitate protein from an aqueous solution. The antimicrobial activities of the naturally occurring phenolics from olives, tea, and coffee have been studied in more detail than those from other sources, which may in part be due to the high value of the products being processed [26]. Phenolics from spices, such as gingerol, zingerone, and capsaicin, have been found to inhibit germination of bacterial spores. Native plant phenolics are important food preservative factors and have, as a group, an impressive antimicrobial spectrum, although their deliberate use as food preservatives is rarely exploited [6].

It is notable that phenolic mixtures can have helpful effects against malignancy, cardiovascular sickness, and numerous different illnesses [27]. The positive effects of phenolic compounds on different ongoing degenerative infections, cardiovascular illnesses, and diseases [28] could be because of their cell reinforcement efficacy, bringing down the measure of free extremists inside the body [27]. Free extremists can make harm cells and their constituents that can prompt the beginning and advancement of ongoing degenerative infections. Also, cell reinforcements hinder the activity of proteins that catalyze the detoxification pathway in the body upon openness to mutagens [29]. In view of their bioactive effects, phenolic mixtures can likewise add to the nature of food.

The metabolites of dietary phenolics in the circulatory system in to low concentrations, exert modulatory effects in cells through selective actions on different components of the intracellular signalling cascades vital for cellular functions such as growth, proliferation, and apoptosis [27].

8.3.5 Essential Oils and Their Components

Essential oils are generally obtained from herbs and spices yet can likewise be segregated from organic products, roots, and stems of plants. A few oils and detached plant compounds are utilized in food as seasoning specialists. These compounds, emanating from their function in plants, show a wide scope of intriguing natural activities [30]. A few mixtures have been appeared to draw in flies for fertilization, while others show a particular creepy crawly repulsing movement. Others pull in herbivores for seed appropriation or show fungicidal or bactericidal action to smother disease by plant pathogenic microorganisms. Mixtures that are endorsed for use in food and consolidated antimicrobial action with low mammalian poisonousness have incredible potential for application as characteristic food additives.

The antimicrobial activity of concentrates obtained from herbs, spices, and other sweet-smelling plants or parts thereof utilizing natural solvents or steam refining has been perceived for a long time. Plants and plant separates have been utilized since relic in society for medication and food safeguarding, giving a scope of mixtures having pharmacological activities [31].

These common mixtures are by and large perceived as safe for climate and human well-being, so interest in their utilization in the objective for practical horticulture has expanded and a ton of exploration has been done, demonstrating, by and large, that plant essential oils and concentrates many have a job as drugs and food conservations [32]. Lately, various investigations have recorded the antifungal impact of plant essential oils to control food deterioration organisms [10, 16, 33–37]. Essential oils of thyme, cinnamon and anise, and mint have more impact on contagious improvement and resulting mycotoxin creation in wheat grains [38].

These mixtures might be deadly to microbial cells or they may restrain the creation of auxiliary metabolites (e.g., mycotoxins) [39]. Plant fundamental oils are by and large more inhibitory against Gram-positive than Gram-negative microorganisms [40, 41]. While this is valid for some fundamental oils, there are a few specialists that are viable against the two gatherings, such as oregano, clove, cinnamon, and citral [40, 42]. The significant fundamental oil parts with antimicrobial impacts found in plants, spices, and flavors are phenolic compounds, terpenes, aliphatic alcohols, aldehydes, ketones, acids, and isoflavonoids [17, 43]. Essential oils have antimicrobial vitalities against a few pathogenic microorganisms present in meat, including both Gram-positive and Gram-negative microbes [33].

Numerous investigations have been led to examine the impacts of essential oils extricated from sources such as oregano, rosemary, thyme, basil, garlic, and clove, when utilized alone or in blend with other fundamental oils [4, 6].

8.4 Mechanisms of Antimicrobial Action

The effect of phenolic compounds as an antimicrobial agent is mentioned above. But even so, the specific instrument of activity is not clear. The impact of phenolic mixtures on its antimicrobial activities [44]. At low focus, phenols influence compound movement, especially those related to energy creation, while at high fixations, they cause protein denaturation. The antimicrobial impact of phenolic mixtures might be because of their capacity to modify microbial cell porousness, in this way allowing the deficiency of macromolecules from the inside (e.g., ribose and Na glutamate) [45]. They could likewise meddle with film work (electron transport, supplement take-up, protein, nuclein corrosive blend, and catalyst movement) [45] and cooperate with layer proteins, causing deformity in construction and usefulness. The high antibacterial movement of phenolic parts can be additionally clarified as far as alkyl replacement into the phenol core [46]. The development of phenoxyl revolutionaries that associate with alkyl substituents does not happen with more steady particles, for example, the ethers myristicin or anethole, which was identified with the overall absence of antimicrobial movement of fennel, nutmeg, or parsley essential oils [27].

Ref. [47] announced that the antimicrobial movement of isothiocyanates obtained from onion and garlic is identified with the inactivation of extracellular catalysts through oxidative cleavage of disulfide bonds and that the development of the receptive thiocyanate extremist was proposed to intercede the antimicrobial impact. Carvacrol, carvone, thymol, and *trans*-cinnamaldehyde are accounted for to diminish the intracellular ATP (adenosine triphosphate) substance of *E. coli* O157:H7 cells while at the same time expanding extracellular ATP, demonstrating the troublesome activity of these mixtures on the plasma layer [48]. Inactivation of yeasts can be credited to the aggravation of a few enzymatic frameworks, such as energy creation and primary part amalgamation [20].

8.5 Natural Antimicrobials of Animal Origin

8.5.1 Chitosan

Chitosan is obtained from halfway deacetylation of chitin and in some cases known as deacetylated chitin. It is a characteristic polycationic straight polysaccharide for the most part and found in shells of marine scavengers [49]. Because of its nontoxicity, biodegradability and also low allergenicity, it has a wide application. It has antitumor, antifungal, and antimicrobial cancer prevention agent activities [37]. It is compelling against Gram-negative microbes such as *Bacteroides fragilis*, cholera, *Shigella dysenteriae*, *E. coli*, and *Vibrio*. Chitosan has great antimicrobial protection from expanding and cell reinforcement potential.

8.5.2 Defensins

Defensins are little cationic peptides and are fundamentally known for their antimicrobial activities that are for the most part antibacterial and antimycotic. They are found in all mammalian cells and tissues and plentiful in leukocytes [50]. There are studies that recorded a wide spectrum of them as antimicrobials, including bacteria (Gram negative and Gram positive), fungi, and viruses.

8.5.3 Lactoferrin

Lactoferrin (LF) is an iron-restricting and bioactive glycoprotein likewise named lactosiderophilin or lactotransferrin. Lactoferrin is available in many body discharges (conceptive, stomach related, and respiratory) like those from the frameworks. It is present in a huge sum in cow-like colostrum than in developed milk. Lactoferrin shows solid antimicrobial impacts against different Gram-negative and Gram-positive microorganisms, organisms, and parasites. It has been appeared to effectively affect changed pathogenic microorganisms including bacteriostatic ones [51].

8.5.4 Lactoperoxidase

Lactoperoxidase (LP) has a place in peroxidase family, and its essential capacity is to catalyze the oxidation of specific particles. It is a gathering of common chemicals, generally dispersed in nature and found in plants and creatures, including man. LP is emitted by ductal epithelial cells of the mammary organ. The level of LP in cow-like milk is around multiple times higher than that of human milk and changes continually during the post pregnancy time frame. Thiocyanate, which is available in critical sums in salivation, milk, and aviation route emission framework, is needed for the antimicrobial movement. Microorganisms including salmonellae, Shigella, pseudomonads, and coliforms are inhibited by LP [52].

8.5.5 Lysozyme

It is a solitary chain polypeptide of 129 amino acids normally present in real discharges such as tears, spits, and milk. It has great antimicrobial adequacy and causes the passing of microscopic organisms by severing a glycosidic linkage of bacterial cell dividers peptidoglycan. Lysozyme is a significant safeguard instrument and is viewed as a piece of the inborn insusceptible framework in many warm-blooded animals [53] and is likewise a significant segment of human bosom milk [54. Williams and Vocadlo 55] reported that in egg white there were a lot of lysozymes.

8.5.6 Avidin

Avidin is an emphatically charged glycoprotein which is available in eggs. Egg too contains biotin. Avidin ties biotin (avidin-biotin framework as a symptomatic device in immunoassays) and makes it inaccessible for the utilization of microorganisms. The action of *E. coli*, *Klebsiella pneumoniae*, *Serratia marcescens*, and *P. aeruginosa* can be controlled [56].

8.5.7 Pleurocidin

Pleurocidin is an antimicrobial peptide composed of 25 amino acids with high antibacterial activity for both Gram-positive and Gram-negative bacteria. It has potential for use in food applications because of its warmth, solidness, salt-lenient capacity, and different qualities. In numerous examinations, it has been discovered to be successful against numerous pathogenic living beings such as *E. coli* O157:H7, *V. parahaemolyticus*, and *L. monocytogenes* [57].

8.6 Mechanism of Antimicrobial Action

The component of activity of AMPs appears to include multiple targets. The plasma membrane is the most referred to target; nonetheless, ongoing investigations propose intracellular focuses at any rate for certain peptides [17, 38]. Albeit most AMPs act by vague instruments, they frequently show some selectivity between different microorganisms, for example, Gram-negative contrasted and Gram-positive microbes [31, 58] and the powerlessness of parasitic cells contrasted and other eukaryotic cells [59]. Antimicrobial peptides can accept amphipathic structures, which can connect straightforwardly with the microbial cell film, quickly upsetting the layer in a few areas, bringing about filtering out of imperative cell parts [59, 60].

Past investigations revealed that the system of activity of pleurocidin displays solid film movement and pore-arrangement capacity, responding with both impartial and acidic anionic phospholipid layers [32]. Lipids inactivate microorganisms principally by disturbance of bacterial cell divider or film, hindrance of intracellular replication, or restraint of an intracellular objective [50]. Monoacylglycerols bring down the warmth opposition of specific microorganisms and parasites; in this way, they may discover application in decreasing the necessary warmth treatment beyond a shadow of food sources (a day and a half). Lysozyme hydrolyses the β-1,4-glycosidic linkage in sugar polymers, for example, *N*-acetylmuramic corrosive and *N*-acetylglucosamine linkages found in bacterial peptidoglycan [31].

8.7 Natural Antimicrobials of Microbial Origin

8.7.1 Lactic Acid Bacteria as Protective Cultures

Lactic acid bacteria are significant probiotics that give numerous medical advantages in remembering defensive parts for food sources. They go about as additive and hinder the development of numerous pathogenic microorganisms. They restrain the development of microorganisms by delivering antimicrobial specialists like natural acids and bacteriocins (antimicrobial peptides) [47].

Different strains of the microbes are viable against numerous microorganisms.

8.7.2 Bacteriocins Produced by Lactic Acid Bacteria

Principle antimicrobial mixtures are created by numerous Gram-positive microorganisms. These are really metabolites of the LAB that are delivered during their development. These polypeptides give the delivering microorganism an upper hand over other kinds of microorganisms. Bacteriocins are ordered depending on their synthetic nature and principally delivered by Gram-positive microscopic organisms [61] (Table 8.1).

Table 8.1 Lactic acid bacteria (LAB) as a source of bacteriocins

Category	Target microorganism	Reference
Lactacin B enterococci	*Lactobacillus acidophilus*, lactobacilli, *Lactococcus lactis*	[7]
Lacticin F	*L. acidophilus*, lactobacilli	[62]
Nisin	*L. lactis* subsp *lactis*	[63]
Lacticin 3147	*L. lactis* subsp *lactis*	

8.8 Conclusion

With the expanding request for new, semi-handled, prepared, and crude food items, their well-being quality and protection issues should be tended to. The guideline and new strategy for use of characteristic antimicrobial specialists are significant elements that ought to be tended to. Streamlining of use techniques, what's more, guideline will upgrade the shopper certainty. The application strategies for the regular antimicrobial specialists to various food items require higher effectiveness. The utilization of regular antimicrobials on leafy foods without damagingly influencing the sensorial qualities is as yet a test for scientists. To restrain the development of waste or dispose of pathogenic microscopic organisms, the necessary convergences of regular antimicrobial specialists are high. These high fixations can influence tactile characteristics as well as contrarily affect human well-being. Exploration should be done on the synergistic mixes of normal antimicrobial specialists. A mix of various medicines emphatically guarantees the security and quality issues of the food items.

References

1. J. Samelis and others, "Managing microbial spoilage in the meat industry," *Food spoilage Microorg.*, pp. 213–286, 2006.
2. E. B. Bingöl and K. Bostan, "Effect of sodium lactate on the microbiological quality and shelf life of sausages," *Turkish J. Vet. Anim. Sci.*, vol. 31, no. 5, pp. 333–339, 2007.
3. K. E. Ricker and R. M. Bostock, "Evidence for release of the elicitor arachidonic acid and its metabolites from sporangia of Phytophthora infestans during infection of potato," *Physiol. Mol. Plant Pathol.*, vol. 41, no. 1, pp. 61–72, 1992.
4. L. Atarés, C. De Jesús, P. Talens, and A. Chiralt, "Characterization of SPI-based edible films incorporated with cinnamon or ginger essential oils," *J. Food Eng.*, vol. 99, no. 3, pp. 384–391, 2010.
5. M. P. Doyle and M. C. Erickson, "Emerging microbiological food safety issues related to meat," *Meat Sci.*, vol. 74, no. 1, pp. 98–112, 2006.
6. Y. Fu *et al.*, "Antimicrobial activity of clove and rosemary essential oils alone and in combination," *Phyther. Res.*, vol. 21, no. 10, pp. 989–994, 2007.
7. P. M. Muriana and T. Klaenhammer, "Purification and partial characterization of lactacin F, a bacteriocin produced by Lactobacillus acidophilus 11088," *Appl. Environ. Microbiol.*, vol. 57, no. 1, pp. 114–121, 1991.
8. M. L. Wells *et al.*, "Algae as nutritional and functional food sources: revisiting our understanding," *J. Appl. Phycol.*, vol. 29, no. 2, pp. 949–982, 2017.
9. K. Sharma, S. Guleria, V. K. Razdan, and V. Babu, "Synergistic antioxidant and antimicrobial activities of essential oils of some selected medicinal plants in combination and with synthetic compounds," *Ind. Crops Prod.*, vol. 154, p. 112569, 2020.
10. H. S. Tabaestani, N. Sedaghat, E. S. Pooya, A. Alipour, and others, "Shelf life improvement and postharvest quality of cherry tomato (Solanum

lycopersicum L.) fruit using basil mucilage edible coating and cumin essential oil," *Int. J. Agron. Plant Prod.*, vol. 4, no. 9, pp. 2346–2353, 2013.

11. I. Roca *et al.*, "The global threat of antimicrobial resistance: science for intervention," *New Microbes New Infect.*, vol. 6, pp. 22–29, 2015.

12. S. A. Sofi, J. Singh, S. Rafiq, U. Ashraf, B. N. Dar, and G. A. Nayik, "A comprehensive review on antimicrobial packaging and its use in food packaging," *Curr. Nutr. Food Sci.*, vol. 14, no. 4, pp. 305–312, 2018.

13. P. Nguyen and G. S. Mittal, "Inactivation of naturally occurring microorganisms in tomato juice using pulsed electric field (PEF) with and without antimicrobials," *Chem. Eng. Process. Process Intensif.*, vol. 46, no. 4, pp. 360–365, 2007.

14. Z. El Astal, "The inhibitory action of aqueous garlic extract on the growth of certain pathogenic bacteria," *Eur. Food Res. Technol.*, vol. 218, no. 5, pp. 460–464, 2004.

15. J. Gutierrez, G. Rodriguez, C. Barry-Ryan, and P. Bourke, "Efficacy of plant essential oils against foodborne pathogens and spoilage bacteria associated with ready-to-eat vegetables: antimicrobial and sensory screening," *J. Food Prot.*, vol. 71, no. 9, pp. 1846–1854, 2008.

16. M. Viuda-Martos, Y. Ruiz-Navajas, J. Fernández-López, and J. A. Perez-Alvarez, "Effect of added citrus fibre and spice essential oils on quality characteristics and shelf-life of mortadella," *Meat Sci.*, vol. 85, no. 3, pp. 568–576, 2010.

17. D. Mucete, A. Borozan, F. Radu, I. Jianu, and E. Alexa, "Researches about the antimicrobial action of some active principles in Armoracia rusticana," 2005.

18. Y. Jiang and Y. Li, "Effects of chitosan coating on postharvest life and quality of longan fruit," *Food Chem.*, vol. 73, no. 2, pp. 139–143, 2001.

19. T. Mangena and N. Y. O. Muyima, "Comparative evaluation of the antimicrobial activities of essential oils of Artemisia afra, Pteronia incana and Rosmarinus officinalis on selected bacteria and yeast strains," *Lett. Appl. Microbiol.*, vol. 28, no. 4, pp. 291–296, 1999.

20. J. B. Harborne, "The role of phytoalexins in natural plant resistance," 1986.

21. C. J. Baker, E. W. Orlandi, and N. M. Mock, "Harpin, an elicitor of the hypersensitive response in tobacco caused by Erwinia amylovora, elicits active oxygen production in suspension cells," *Plant Physiol.*, vol. 102, no. 4, pp. 1341–1344, 1993.

22. M. W. Örner and H. C. Jha, "Antifungal activity of flavonoids and their mixtures against different fungi occurring on grain," *Pestic. Sci.*, vol. 38, no. 4, pp. 347–351, 1993.

23. H. Dörnenburg and D. Knorr, "Generation of colors and flavors in plant cell and tissue cultures," *CRC. Crit. Rev. Plant Sci.*, vol. 15, no. 2, pp. 141–168, 1996.

24. A. Arnoldi and L. Merlini, "Lipophilicity-antifungal activity relationships for some isoflavonoid phytoalexins," *J. Agric. Food Chem.*, vol. 38, no. 3, pp. 834–838, 1990.

25. A. Stintzi *et al.*, "Plant 'pathogenesis-related' proteins and their role in defense against pathogens," *Biochimie*, vol. 75, no. 8, pp. 687–706, 1993.

26. G. J. E. Nychas, "Natural antimicrobials from plants," in *New Methods of Food Preservation*, Springer, 1995, pp. 58–89.

27. A. Crozier, I. B. Jaganath, and M. N. Clifford, "Dietary phenolics: chemistry, bioavailability and effects on health," *Nat. Prod. Rep.*, vol. 26, no. 8, pp. 1001–1043, 2009.

28. F. Yan, S. Xu, Y. Chen, and X. Zheng, "Effect of rhamnolipids on Rhodotorula glutinis biocontrol of Alternaria alternata infection in cherry tomato fruit," *Postharvest Biol. Technol.*, vol. 97, pp. 32–35, 2014.

29. N. F. Santos-Sánchez, R. Salas-Coronado, C. Villanueva-Cañongo, and B. Hernández-Carlos, *Antioxidant Compounds and Their Antioxidant Mechanism*, IntechOpen London, UK, 2019.

30. N. Nakatani, "Antioxidative and antimicrobial constituents of herbs and spices," in Charalambous G. (ed), *Spices, Herbs, and Edible Fungi*, Elsevier, Amsterdam, 1994.

31. S. G. Deans, K. P. Svoboda, and others, "Biotechnology and bioactivity of culinary and medicinal plants," *AgBiotech News Inf.*, vol. 2, no. 2, pp. 211–216, 1990.

32. K. Yoshida *et al.*, "Interaction of pleurocidin and its analogs with phospholipid membrane and their antibacterial activity," *J. Pept. Res.*, vol. 57, no. 2, pp. 119–126, 2001.

33. H. Shirzad, A. Hassani, Y. Ghosta, A. Abdollahi, R. Finidokht, and M. Meshkatalsadat, "Assessment of the antifungal activity of natural compounds to reduce postharvest gray mould (Botrytis cinerea pers.: fr.) of kiwifruits (actinidia deliciosa) during storage," *J. Plant Prot. Res.*, 2011.

34. J. Tian, X. Ban, H. Zeng, B. Huang, J. He, and Y. Wang, "In vitro and in vivo activity of essential oil from dill (Anethum graveolens L.) against fungal spoilage of cherry tomatoes," *Food Control*, vol. 22, no. 12, pp. 1992–1999, 2011.

35. Y. G. Xing, X. H. Li, Q. L. Xu, C. X. Shao, and J. Yun, "Antimicrobial activity of microencapsulated cinnamon oil and its application on cherry tomato," in *Advanced Materials Research*, 2011, vol. 236, pp. 2307–2310.

36. Q. Chen *et al.*, "Effect of citronella essential oil on the inhibition of postharvest Alternaria alternata in cherry tomato," *J. Sci. Food Agric.*, vol. 94, no. 12, pp. 2441–2447, 2014.

37. D.-H. Ngo and S.-K. Kim, "Antioxidant effects of chitin, chitosan, and their derivatives," *Adv. Food Nutr. Res.*, vol. 73, pp. 15–31, 2014.

38. K. M. Soliman and R. I. Badeaa, "Effect of oil extracted from some medicinal plants on different mycotoxigenic fungi," *Food Chem. Toxicol.*, vol. 40, no. 11, pp. 1669–1675, 2002.

39. P. M. Davidson, "Chemical preservatives and naturally antimicrobial compounds cit Celikel, N. and Kavas, G., 2008, Antimicrobial properties of some essential oils against some pathogenic microorganisms," *Czech J. Food Sci*, vol. 26, no. 3, pp. 174–181, 2001.

40. P. Skandamis, E. Tsigarida, and G. J. E. Nychas, "The effect of oregano essential oil on survival/death of Salmonella typhimurium in meat stored at 5 C under aerobic, VP/MAP conditions," *Food Microbiol.*, vol. 19, no. 1, pp. 97–103, 2002.

41. N. Chorianopoulos, E. Kalpoutzakis, N. Aligiannis, S. Mitaku, G.-J. Nychas, and S. A. Haroutounian, "Essential oils of Satureja, Origanum, and Thymus species: chemical composition and antibacterial activities against foodborne pathogens," *J. Agric. Food Chem.*, vol. 52, no. 26, pp. 8261–8267, 2004.

42. S. Kim and D. Y. C. Fung, "Antibacterial effect of water-soluble arrowroot (Puerariae radix) tea extracts on foodborne pathogens in ground beef and mushroom soup," *J. Food Prot.*, vol. 67, no. 9, pp. 1953–1956, 2004.

43. N. G. Chorianopoulos, E. D. Giaouris, P. N. Skandamis, S. A. Haroutounian, and G.-J. Nychas, "Disinfectant test against monoculture and mixed-culture biofilms composed of technological, spoilage and pathogenic bacteria: bactericidal effect of essential oil and hydrosol of Satureja thymbra and comparison with standard acid–base sanitizers," *J. Appl. Microbiol.*, vol. 104, no. 6, pp. 1586–1596, 2008.

44. H. Dörnenburg and D. Knorr, "Elicitation of chitinases and anthraquinones in Morinda citrifolia cell cultures," 1994.

45. V. K. Bajpai, A. Rahman, N. T. Dung, M. K. Huh, and S. C. Kang, "In vitro inhibition of food spoilage and foodborne pathogenic bacteria by essential oil and leaf extracts of Magnolia liliflora Desr.," *J. Food Sci.*, vol. 73, no. 6, pp. M314–M320, 2008.

46. H.-D. Dorman and S. G. Deans, "Antimicrobial agents from plants: antibacterial activity of plant volatile oils," *J. Appl. Microbiol.*, vol. 88, no. 2, pp. 308–316, 2000.

47. C. E. Isaacs, R. E. Litov, and H. Thormar, "Antimicrobial activity of lipids added to human milk, infant formula, and bovine milk," *J. Nutr. Biochem.*, vol. 6, no. 7, pp. 362–366, 1995.

48. I. M. Helander, H.-L. Alakomi, K. Latva-Kala, and others, "Characterization of the action of selected essential oil components on Gramnegative bacteria," *J. Agric. Food Chem.*, vol. 46, pp. 3590–3595, 1998.

49. T. Chandy and C. P. Sharma, "Chitosan-as a biomaterial," *Biomater. Artif. Cells Artif. Organs*, vol. 18, no. 1, pp. 1–24, 1990.

50. B. Agerberth *et al.*, "The human antimicrobial and chemotactic peptides LL-37 and α-defensins are expressed by specific lymphocyte and monocyte populations," *Blood J. Am. Soc. Hematol.*, vol. 96, no. 9, pp. 3086–3093, 2000.

51. A. A. Al-Nabulsi and R. A. Holley, "Effect of bovine lactoferrin against Carnobacterium viridans," *Food Microbiol.*, vol. 22, no. 2–3, pp. 179–187, 2005.

52. B. Reiter and G. Härnulv, "Lactoperoxidase antibacterial system: natural occurrence, biological functions and practical applications," *J. Food Prot.*, vol. 47, no. 9, pp. 724–732, 1984.

53. S. Varahan, V. S. Iyer, W. T. Moore, and L. E. Hancock, "Eep confers lysozyme resistance to Enterococcus faecalis via the activation of the extracytoplasmic function sigma factor SigV," *J. Bacteriol.*, vol. 195, no. 14, pp. 3125–3134, 2013.

54. B. Lönnerdal, "Nutritional and physiologic significance of human milk proteins," *Am. J. Clin. Nutr.*, vol. 77, no. 6, pp. 1537S–1543S, 2003.

55. S. Williams and D. Vocadlo, "Glycoside hydrolase family 22," *Cazypedia. Retrieved*, vol. 11, 2017.

56. J. Korpela, "Avidin, a high affinity biotin-binding protein, as a tool and subject of biological research," *Med. Biol.*, vol. 62, no. 1, pp. 5–26, 1984.

57. O. J. Burrowes, C. Hadjicharalambous, G. Diamond, and T.-C. LEE, "Evaluation of antimicrobial spectrum and cytotoxic activity of pleurocidin for food applications," *J. Food Sci.*, vol. 69, no. 3, pp. FMS66–FMS71, 2004.

58. I. Helander, A. von Wright, and T. M. Mattila-Sandholm, "Potential of lactic acid bacteria and novel antimicrobials against Gram-negative bacteria," *Trends Food Sci. Technol.*, vol. 8, no. 5, pp. 146–150, 1997.

59. R. E. W. Hancock, "Peptide antibiotics," *Lancet*, vol. 349, no. 9049, pp. 418–422, 1997.

60. A. M. Cole, R. O. Darouiche, D. Legarda, N. Connell, and G. Diamond, "Characterization of a fish antimicrobial peptide: gene expression, subcellular localization, and spectrum of activity," *Antimicrob. Agents Chemother.*, vol. 44, no. 8, pp. 2039–2045, 2000.

61. S. C. Ricke, "Anaerobic microbiology laboratory training and writing comprehension for food safety education," in *Food Safety*, Elsevier, 2015, pp. 395–419.

62. A. Hurst, "Nisin," in *Advances in Applied Microbiology*, vol. 27, Elsevier, 1981, pp. 85–123.

63. O. McAuliffe, M. P. Ryan, R. P. Ross, C. Hill, P. Breeuwer, and T. Abee, "Lacticin 3147, a broad-spectrum bacteriocin which selectively dissipates the membrane potential," *Appl. Environ. Microbiol.*, vol. 64, no. 2, pp. 439–445, 1998.

Chapter 9

Plant-Based Food Preservatives

Yeliz Tekgül,[a] Edibe S. Erten,[b] and Gülşah Ç. Koç[c]
[a]*Aydın Adnan Menderes University, Köşk Vocational High School, Food Processing Department, Aydın, Turkey*
[b]*Aydın Adnan Menderes University, Engineering Faculty, Food Engineering Department, Aydın, Turkey*
[c]*Uşak University, Eşme Vocational High School, Food Technology Program, Eşme, Uşak, Turkey*
gulsah.koc@usak.edu.tr

Food preservatives are used for delaying the degradation of foods because of oxidation, microbial growth, and enzyme activity. The tendency of using natural additives is increasing due to the negative health effects of synthetic preservatives. The phytochemicals such as polyphenols, flavonoids, tannins, alkaloids, terpenoids, lectins, polypeptides, and so on in the plant (herbs, spices, etc.) extracts and essential oils provide antimicrobial activity against yeast, bacteria, molds, viruses, and insects and antioxidant activity against oxidative degradation. For this reason, plant extracts and essential oils are generally used as food preservatives. More than 17,000 aromatic

Natural Food Preservatives
Edited by Sourish Bhattacharya and Heba Hassan Abd-El Azim Salama

ISBN 978-981-4968-64-5 (Hardcover), 978-1-003-36776-5 (eBook)
www.jennystanford.com

plant species such as Lamiaceae, Rutaceae, Myrtaceae, and so on are used for obtaining essential oil. Plant extracts are obtained from plant parts such as leaves, stems, flowers, fruits, roots, and so on. To discover new natural plant-based preservatives and extraction methods, identify their active ingredients, and investigate their effect on the foods have to be focused further studies.

9.1 Introduction

For a long time, various food ingredients such as spices have been used for keeping the shelf life of foods longer to consume later. With the progress in food science, numerous compounds from other foods have been found to have either antimicrobial or antioxidant effects or both. Although most of these studies were conducted for medical and health aspects, the food industry has also been benefited. In recent years, the tendency to have natural preservatives instead of synthetic chemicals has become a trend for consumers. Furthermore, for the purpose of reducing the waste in the food industry, rejected parts of food plants such as peels, seeds, and skins have been studied for their antioxidant and antimicrobial effects. These parts of plants have found their applications in foods as either powdered form or as extracts that are generally applied to packaging. The limitations of the usage of plant-originated preservatives in food products are their potential toxicity at high concentrations and the prolonged period of time and research requirements for their regulatory acceptance.

For the preservation effects, the food has been used directly as an ingredient or the extract of the food or isolated compounds with preservative effects have been added to food products. The application of extracts in edible coating or packaging material is also another method for usage. As a means of extraction, there are traditional methods, such as distillation with steam or organic solvents, and novel methods such as high vacuum and high-pressure extractions, supercritical fluid extraction, solvent-free by microwave, and so on that have been practiced [1]. The extraction steps and extraction methods are shown in Fig. 9.1.

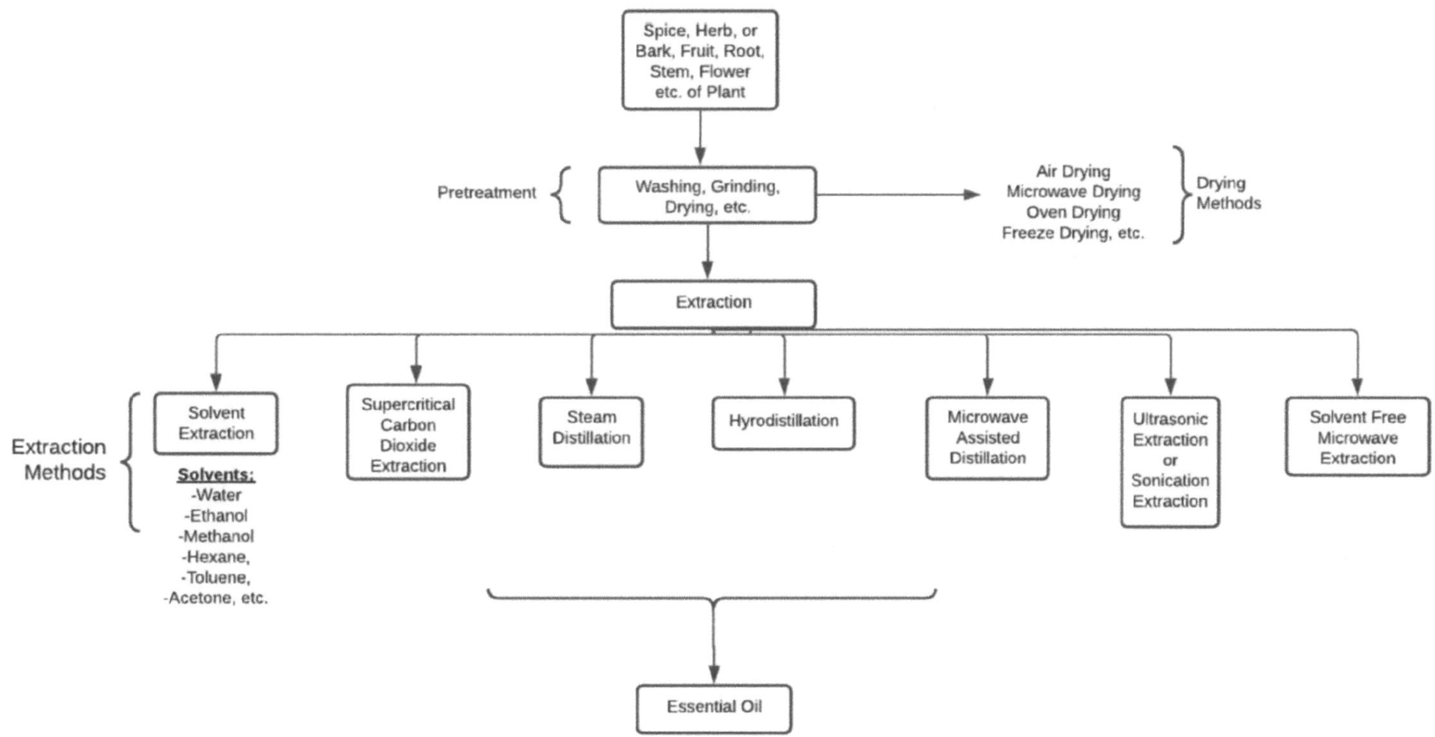

Figure 9.1 The extraction steps and extraction methods.

Numerous studies have been conducted to determine the antimicrobial and antioxidant properties of edible plants and their components [2, 3]. The antimicrobial effect of products is usually analyzed by inoculating the target microorganism to the food products along with the intended foods, their extracts, or essential oils. After incubation, the final count of microorganisms is compared with the initial count. Antioxidant activity of foods is analyzed by several techniques such as the Folin-Ciocalteu method, oxygen radical absorbance capacity (ORAC), DPPH (2,2-diphenyl-1-picrylhydrazyl) free radical scavenging, and total radical trapping antioxidant parameter (TRAP) assay methods. Antimicrobial properties of foods are originated from their phenolic content, whereas antioxidant activities of food products are generally associated with their bioactive and/or phenolic components, such as alkaloids, polyphenols, flavonoids, terpenoids, and isothiocyanates [1]. In this chapter, the food preservation properties of natural foods, their extracts, and their by-products are discussed from the aspect of recent findings.

9.2 Using Spices as Food Preservatives

Spices and natural herbs have preservative activity depending on the type, nature, and concentration of the organism. They contain volatile compounds used in preservative production. Compounds found in spices and natural herbs have a protective role to play in the plant against infectious organisms as self-defense mechanisms. Plant oils and herbal extracts have been used as food preservatives and medicinal treatment agents for many years. The essential oil, phenolic compounds, and extracts of plants and herbs which are usually obtained from their leaves, flowers, roots, stems, barks, and fruits are antimicrobial components. When the literature was examined, it was seen that many spices had antioxidant and antimicrobial effects. Within the scope of this chapter, the studies investigating the antimicrobial or antioxidant effects of extracts or essential oils of spices by applying them directly to foods or packaging materials are included (Table 9.1). Although many spices and herbs such as laurel, eucalyptus, basil, anise, black pepper, sumac, clove, and ginger have been included in the studies, the antimicrobial and antioxidant activities of the frequently used spices and herbs such as red pepper, cinnamon, mint, and so on are examined in this chapter.

9.2.1 Red Pepper

Red peppers are generally consumed as fresh, but it is also used as a spice and colorant by pulverizing. Red pepper is rich in ascorbic acid, polyphenols (tocopherol, β-carotene, etc.), and so on which contribute to its antioxidant and antimicrobial activities. Capsaicin—the bitter substance in red peppers—has a protective effect against cancer and cardiovascular diseases. Capsanthin, which gives color to red peppers, is polyphenolic like capsaicin that also has antioxidant properties. Since red pepper is effective against both Gram-positive (+) and Gram-negative (–) bacteria, it can be evaluated as a natural food preservative against foodborne pathogens in the fresh or dry form [4]. The addition of 3% pepper in the production of chorizo, a type of sausage unique to Spain, has been reported to have antioxidant activities such as nitrite, nitrate, and ascorbic acid [5]. On the other hand, the use of pepper in the production of chorizo causes a decrease in biogenic amines. The addition of red pepper to beef patties has demonstrated antimicrobial activity against *Listeria monocytogenes* and *Micrococcus luteus*. In addition, the thiobarbituric acid reactive substance (TBARS) values of beef meatballs which include 0.35% red pepper were found to be lower at the end of 7 days compared to the samples using synthetic antioxidants. The use of red pepper in the production of Kareish cheese has an antimicrobial effect against coliforms, yeast, molds, and *Staphylococcus aureus* strains [6].

9.2.2 Caraway

Caraway has several health beneficial effects due to its anti-obesity, analgesic, anti-inflammatory, antianxiety, and antispasmodic properties. Carvacrol, carvone, α-pinene, limonene, γ-terpinene, carvenone, linalool, and p-cymene are the major volatile components of caraway. Caraway is used in food products as well as in therapeutics. Essential oils obtained from this plant show antimicrobial activities. The addition of caraway extract to the silver carp extended the shelf life of the product in refrigerator conditions by 6–9 days [7]. The use of essential oils obtained from this plant in Iranian white cheese has inhibited pathogens [8]. In a study in which the effect of active biobased film application containing caraway oil on the shelf life of nuggets was examined, it was found that the biobased film application was effective against *Bacillus cereus*, *Escherichia coli*, *Pseudomonas aeruginosa*, and *S. aureus*. This also extended the shelf life of the nuggets for 15 days of samples stored at +4°C [9].

9.2.3 Cinnamon

There are 250 different types of cinnamon grown around the world. Ceylon (*Cinnamon zeylanicum Blume*), Vietnamese (*Cinnamon loureiroi Nees*), Indonesian (*Cinnamon burmanni* (*Nees & T. Nees*)), and Chinese (*Cinnamon cassia* (L.). J. Presl) cinnamons are economical and widely consumed. In addition to being consumed as a spice, the essential oils obtained from the peel, leaves, flowers, and fruit are also used in medicine, perfumery, cosmetics, and food industries. It is preferred as a preservative additive in various foods, especially due to its strong antimicrobial effect. (*E*)-cinnamaldehyde, benzaldehyde, (*E*)-cinnamyl acetate, limonene, and eugenol are the major volatile components found in cinnamon. Vanillic, caffeic, gallic, protocatechuic, *p*-coumaric, and ferulic acids are among the main polyphenols found in cinnamon. As a result of enrichment with cinnamon oil in coating with chitosan, the shelf life of trout was increased from 12 to 16 days [10]. The use of agar and alginate bilayer films containing cinnamon oil, which has antioxidant and antimicrobial properties, in shrimp decreased the number of pathogenic microorganisms in the samples [11].

9.2.4 Cumin

It can be used as a spice to the flavoring of foods, as well as to treat diseases such as epilepsy, dyspepsia, and jaundice. Components such as γ-terpinene, *p*-cymene, pinene, cumin aldehyde, safranal, and cuminal found in cumin oil show antioxidant and antimicrobial activities. Cuminal and cuminic alcohol are strong antioxidants that remove hydroxyl radicals, 1,1-diphenyl-2-picrylhydrazyl (DPPH) radicals, and lipid peroxidase. Phenolics and essential oils in cumin can be used as natural antioxidants in the food industry. The use of cumin oil at 0.1% and 0.2% (weight/weight, w/w) concentrations in cheese production decreases the growth of pathogenic microorganisms [12]. To eliminate undesirable changes in foods, cumin essential oils are used in active packaging based on the inclusion of preferred natural antimicrobial substances into packaging materials. The shelf life of cultivated mushrooms packaged with nano biopolymers containing cumin oil is extended [13]. It has been determined that packaging with active polyethylene terephthalate (PET) films combined with antimicrobial chitosan and alginate coatings containing cumin oil is effectively protecting the quality and safety of chicken meats [14]. The use of active papers containing encapsulated cumin oil in beef hamburger packaging extends the shelf life of the products [15].

9.2.5 Fennel

Fennel shows anti-inflammatory, anticarcinogenic, antidiabetic, hepatoprotective, antinociceptive, hypolipidemic, antihirsutism, antithrombotic, and antidepressant activities. A significant part of the essential oils obtained from its fruits constitutes anethole and fenchone. The use of fennel in cottage cheese production prevents microbial growth, makes the product safe, and extends the shelf life of the product [16]. The shelf life has increased to over 18 days of fish fillets coated with chitosan nanoparticles (1%, volume/volume, v/v), to which fennel essential oil was added (1%, v/v) and packaged in a modified atmosphere [17]. The use of fennel seeds in biscuit production reduces the formation of acrylamide in products [18]. Tomatoes coated with guar gum containing fennel essential oil have a shelf life of over 60 days [19].

9.2.6 Mint

Lamiaceae consists of more than 7000 species. *Mentha arvensis* L. (corn mint), *Mentha × piperita* L. (peppermint), *M. citrata* Ehrh. (bergamot mint), *M. longifolia* L. (wild mint), and *M. spicata* L. (spearmint) are widely consumed. Menthol, menthone, menthofuran, isomenthone, (E)-caryophyllene, 1,8-cineole, linalool, limonene, carvone, pulegone, and α-terpineol are the major ingredients found in the essential oils of mint species. Peppermint oil is encapsulated and used in yogurt production. Yogurts enriched with peppermint oil have a comparatively longer shelf life [20]. Enrichment with freeze-dried mint decreases the acrylamide formation in bread during baking [21]. The use of *M. spicata* L. and *M. villosa* Huds essential oils in cashew, guava, mango, and pineapple juices effectively controls the spoilage caused by yeasts in fruit juices [22]. The addition of peppermint essential oil in cooked pork sausages reduces the amount of sodium nitrite used in the products [23].

9.2.7 Rosemary

Rosemary officinalis L., *Rosemary erriocaly* Mill., *Rosemary laxiflorus*, and *Rosemary lavandulaceus* are the most common rosemary species. 1.8-cineol, camphor, and borneol are major components in essential oils obtained from the leaves of rosemary. Carnosic acid, carnosol, and rosmarinic acid are the main diterpenes found in rosemary that are reported for their antioxidant activities. Pectin coating enriched with rosemary oil showed high antimicrobial activity on broccoli flowers [24]. The use of rosemary extract in the production of chocolate muffins has shown similar effects with potassium sorbate, a synthetic preservative [25]. The coating process with the film obtained from whey protein concentrate supplemented with rosemary oil has extended the shelf life of spinach and increased its quality [26].

9.2.8 Thyme

In addition to being used as a flavoring agent in foods and beverages, thyme is also used as an antioxidant and antimicrobial additive since it contains components such as thymol, carvacrol, and cymene in its essential oil. Chitosan-gelatin nanofibers containing thyme oil can be used as a substitute for nitrite in sauces [27]. The use of *Thymus serpyllum* L. flour extract, a wild thyme variety in pork patties, prevented lipid and protein oxidations in the products and extended the shelf life of the products [23]. The use of encapsulated *T. vulgaris* essential oils in cakes has increased the shelf life to 30 days without the use of synthetic [28]. The use of thymol essential oil with oregano in tomato juices has created synergistic antimicrobial activity against *Leuconostoc citreum* [29].

Table 9.1 Preservation effects of selected spices

Product	Effect	Effect on	Ref.
***Capsicum annum* L. (red pepper)**	Antioxidant activity Antimicrobial properties	*S. aureus* *P. aeruginosa* *E. coli* *Salmonella typhimurium*	[5]
***C. frutescens* (red pepper)**	Antimicrobial properties	*S. aureus* *E. coli* Yeast and mold	[6]
***Carum carvi* L. (caraway)**	Antioxidant activity Antimicrobial properties	Total viable aerobic bacteria Psychrotrophic bacteria	[7]
***C. carvi* L. (caraway)**	Antimicrobial properties	*S. typhimurium* *E. coli* *S. aureus* *L. monocytogenes*	[8]
***C. carvi* L. (caraway)**	Antioxidant activity Antimicrobial properties	*B. cereus* *E. coli* *P. aeruginosa* *S. aureus*	[9]
***Cinnamomum zeylanicum* (L.) (cinnamon)**	Antioxidant activity Antimicrobial properties	Total viable aerobic bacteria Psychrotrophic bacteria	[10]
***C. zeylanicum* (cinnamon)**	Antioxidant activity Antimicrobial properties	*Photobacterium phosphoreum* *S. aureus* *S. choleraesuis* *E. coli* *P. fluorescens* *L. monocytogenes* *Vibrio parahaemolyticus* *Shewanella putrefaciens*	[11]
***Cuminum cyminum* (cumin)**	Antimicrobial properties	*S. aureus* *S. enteritidis* *E. coli* *L. monocytogenes*	[12]

(*Continued*)

Table 9.1 (*Continued*)

Product	Effect	Effect on	Ref.
***C. cyminum* (cumin)**	Antioxidant activity		[13]
***C. cyminum* (cumin)**	Antimicrobial properties	*S. aureus* *E. coli*	[14]
***C. cyminum* (cumin)**	Antioxidant activity Antimicrobial properties	*S. aureus* *E. coli*	[15]
***Foeniculum vulgare* Mill (fennel)**	Antioxidant activity Antimicrobial properties	*E. coli* *P. aeruginosa* *S. typhimurium* *Enterobacter cloacae* *S. aureus* *B. cereus* *M. flavus* *L. monocytogenes* *Aspergillus fumigatus* *A. ochraceus* *A. versicolor* *A. niger* *Penicillium funiculosum* *P. ochrochloron*	[16]
***F. vulgare* Mill (fennel)**	Antioxidant activity Antimicrobial properties	*Pseudomonas*	[17]
***F. vulgare* L. (fennel)**	Antioxidant activity		[18]
***F. vulgare* Miller (fennel)**	Antioxidant activity Antimicrobial properties	Total viable counts	[19]

Product	Effect	Effect on	Ref.
***M. spicata* L. and *M. villosa* Huds. (mint)**	Antimicrobial properties	*Candida albicans* *C. tropicalis* *Pichia anomala* *Saccharomyces cerevisiae*	[22]
***Rosmarinus officinalis* L. (rosemary)**	Antioxidant activity Antimicrobial properties	*E. coli* *S. choleraesuis* *L. monocytogenes* *S. aureus*	[24]
***R. officinalis* L. (rosemary)**	Antioxidant activity Antimicrobial properties	*B. cereus* *L. monocytogenes* *S. typhimurium* *E. cloacae* *A. fumigatus* *A. niger* *A. versicolor* *P. funiculosum* *Trichoderma viride* *P. verrucosum*	[25]
***R. officinalis* L. (rosemary)**	Antimicrobial properties	*E. coli*	[26]
***Zataria multiflora* Boiss. (thyme)**	Antimicrobial properties	*Clostridium perfringens*	[27]
***T. serpyllum* L. (thyme)**	Antioxidant activity Antimicrobial properties	Enterobacteriaceae Lactic acid bacteria	[23]
***T. vulgaris* (thyme)**	Antimicrobial properties	*Enterococcus faecium* *E. hirae* *C. albicans* *S. aureus* *A. niger* *E. coli* *S. choleraesuis* *S. typhimurium* *P. aeruginosa*	[28]
***T. vulgaris* (thyme)**	Antimicrobial properties	*L. citreum*	[29]

9.3 Using Fruits, Vegetables, and Other Edible Plants as Food Preservatives

To utilize their preservation properties, in general fruits and fruit-based products find their applications in sweet products such as protein bars, beverages, and bakery products. On the other hand, vegetables are used both in sweet and savory products such as soup mixes and "healthy" snacks. In recent years, studies on the preservation effects of fruits and vegetables have increased compared to past decades (Table 9.2). Berries are one of the most studied fruit groups since they are rich in phenolic compounds and have wide applications in the food industry for their wide spectrum of flavor and taste. Extracts from cranberries, bilberries, blueberries, blackberries, raspberries, and strawberries have been found to have inhibition effects on various pathogens [30, 31]. In a recent study, antioxidant activities of nine small berries (black, white and red currant, blackberries, blueberries, goji, raspberries, white and red gooseberries) were analyzed, and as a result, blueberries and black currants had the highest values [32]. Strawberries have been found to be a rich source of bioactive components with antioxidant activity in both fruit and other parts of the plant [33].

Another widely researched fruit group is citrus varieties and their by-products such as pith, seeds albedo, and flavedo. Especially essential oils from peels have been found to have both antioxidant and antimicrobial effects on plants [34, 35]. Main phenolic compounds of citrus peels that show antioxidant activity were phenolic acids, flavanones, and polymethoxylated flavones [36]. Kumquat and mandarin peel extracts were effective on pathogens and also showed antioxidant activity due to their ascorbic acid and polyphenol contents [37, 38].

Consumption of tropical fruits has been widespread around the world with advances in agricultural practices and logistics. Therefore, increased usage resulted in more studies that were focused on the preservation effects of these plants (both fruits and other parts such as peels and barks). For example, peel and seed extracts of kiwi fruit, dekopon, sapodillo, tamarillo, and noni fruits were analyzed for their antimicrobial activities, and extracts from noni fruits were the most effective sample [39]. Fruit pomace from acerola, guava, and cashew showed antioxidant activity; however, they did not have antimicrobial properties [40]. Banana pulp has also shown antioxidant property with its phenolic content and found application in the food industry as both fresh fruit and banana flour [41]. Hanani et al. [42] applied pomegranate powder on fish gelatin film and observed antioxidant activity and antimicrobial effect on main pathogens. Furthermore, peel, core, stem, and leaf of pineapples and sugar syrup of coconuts were found to have antioxidant activity [43, 44].

Table 9.2 Preservation effects of selected fruits, vegetables, and other edible plant materials

Product	Effect	Effect on	Ref.
***Actinidia deliciosa* (kiwi fruit)**	Antimicrobial activity from peel extract Antioxidant activity of fruit	*S. enteritidis* *S. aureus*	[39] [45]
***Allium cepa* L. (onion)**	Antioxidant activity Antimicrobial properties	*E. coli* *P. auruginasa* *S. typhi* *Streptococcus pyogenes*	[48]
***A. sativum* (garlic)**	Antioxidant activity Antimicrobial properties	*E. coli* *P. aeruginosa* *S. typhi* *S. pyogenes*	[48]
***A. sativum* (garlic)**	Antioxidant activity Antimicrobial properties	*P. fluorescens* *Myxococcus piscicola* *V. anguillarum* *Aeromonas punctata* *F. intestinalis*	[58]
***Ananas comosus* (L) Merr. (pineapple) Peel, core, stem, and leaf**	Antioxidant activity		[43]
***Beta vulgaris* (beet)**	Antimicrobial properties Antioxidant activity	*B. subtilis* *B. cereus* *E. coli*	[49] [50]

(Continued)

Table 9.2 (*Continued*)

Product	Effect	Effect on	Ref.
Brassica	Antioxidant activity Antimicrobial properties	*Escherichia* *Klebsiella* *Salmonella* *Bacillus* *Serratia* *Staphylococcus* *Listeria*	[51]
C. sinensis **(tea)**	Antioxidant activity Antimicrobial properties	*E. coli* *P. aeruginosa* *S. typhi* *S. pyogenes*	[48] [55]
Carica papaya **(papaya)**	Antioxidant activity Antimicrobial properties	*E. coli* *P. aeruginosa* *S. typhi* *S. pyogenes*	[48]
Citrus japonica thumb **(kumquat)**	Antimicrobial and antioxidant activity from peel extract	*S. aureus* *S. typhimurium* *E. coli* *P. aeruginosa*	[38]
C. reticulata **(mandarin)**	Antioxidant activity Antimicrobial properties	*E. faecalis* *L. monocytogenes* *S. aureus* *E. coli* *C. albicans*	[37]
Cocos nucifera **(coconut) sugar syrup**	Antioxidant activity		[44]
Ficus carica **L. (fig)**	Antioxidant activity Antimicrobial properties	*B. subtilis* *C. perfringens* *E. faecalis* *S. aureus* *V. cholerae* *P. aeruginosa* *E. coli* *E. sakazakii* *E. cloacae* *Proteus mirabilis* *Citrobacter freundii* *K. oxytoca* *S. odorifera*	[56]

Product	Effect	Effect on	Ref.
***Fragaria x ananassa* Duch. (strawberry)**	Antimicrobial properties Antioxidant activity of fruit, leaves, roots, and stolon	*S. typhimurium* *Campylobacter jejuni* *C. perfringens* *Helicobacter pylori*	[30] [33]
***Lycium barbarum* L. (goji berry)** ***Ribes grossularia* L. (gooseberry)** ***R. nigrum* (black currant)** ***R. pallidum* (white currant)** ***R. rubrum* (red currant)**	Antioxidant activity		[32]
***L. barbarum* L. (goji berry)**	Antioxidant activity Antimicrobial properties	*E. coli* *S. typhimurium* *C. jejuni* *S. aureus* *L. monocytogenes* *C. perfringens* *Yarrowia lipolytica* *Metschnikowia fructicola* *Rhodotorula mucilaginosa* *P. expansum* *A. niger* *Fusarium oxysporum* *Rhizoctonia solani*	[59]
***Musa × paradisiaca* L. (banana)**	Antioxidant activity from pulp		[41]
***Pistacia atlantica* (mastic gum)**	Antioxidant activity and antimicrobial properties from essential oil	*E. coli* *P. aeruginosa* *S. typhimurium* *S. faecalis* *B. cereus* *S. aureus* *C. albicans*	[54]

(*Continued*)

Table 9.2 (*Continued*)

Product	Effect	Effect on	Ref.
Pomegranate	Antioxidant activity and antimicrobial property from skin powder	*B. cereus* *S. aureus* *S. aureus* *L. innocua* *L. monocytogenes* *S. enteritidis* *P. aeruginosa* *E. coli* *Lactobacillus plantarum* *L. rhamnosus* *L. acidophilus* KI *Bifidobacterium animalis* BB12 *B. animalis* BO	[45] [60]
***Psidium guajava* (guava)**	Antioxidant activity from pomace		[40] [45]
***Punica granatum* (pomegranate)**		*L. monocytogenes* *S. aureus* *E. coli*	[42]
***Rubus fruticosus* (blackberry)**	Antioxidant activity Antimicrobial properties	*H. pylori* *Porphyromonas gingivalis* *Fusobacterium nucleatum* *Streptococcus* spp. *E. coli* O157:H7 *L. monocytogenes*	[30] [32]
***R. idaeus* (raspberries)**	Antioxidant activity Antimicrobial properties	*S. typhimurium* *B. subtilis* *S. aureus*	[30] [32]

Product	Effect	Effect on	Ref.
***Vaccinium corymbosum* (blueberries)**	Antioxidant activity Antimicrobial effect from fruits Antimicrobial effect from peels	*L. monocytogenes* *S. enteritidis* *S. typhimurium* *C. jejuni* *E. coli* O157:H7 *L. innocua* *E. faecalis* *B. cereus* *P. aeruginosa* *V. parahaemolyticus* Gram-negatives	[45] [30] [32] [31]
***V. oxycoccos* (cranberry)**	Antioxidant activity Antimicrobial properties	*P. aeruginosa* *L. monocytogenes* *E. coli* O157:H7 *S. typhimurium* *S. aureus*	[45] [30]
***V. uliginosum* (Chinese blueberry)**	Antimicrobial properties (anthocyanins)	*L. monocytogenes* *S. aureus* *S. enteritidis* *V. parahaemolyticus*	[57]
***Vitis vinifera* (grape)**	Antioxidant activity and antimicrobial properties from seeds and skins Antioxidant activity from pomace and seed extract	*Bacillus* sp. *Lactococcus* sp. *Lactobacillus* sp. *Leuconostoc* sp. *Micrococcus* sp. *Streptococcus* sp.	[46] [37]
Whole grain cereals (wheat, barley, oat, millet)	Antioxidant activity		[52]

Another group of fruits that shows preservation activity are orchard fruits such as apples, apricots, peaches, and pears. Extensive research has been conducted on these fruits since they are well known around the globe for a long time and they show antioxidant activity with their content of phenolic compounds [45]. There are also recent studies conducted on grapes as vine fruit. In addition to antioxidant activity from grape pomace, both antioxidant and antimicrobial effects were detected in phytochemical extracts of seeds and skins of grapes [46, 37].

Compared to fruits, vegetables generally are sturdier than fruits and have a longer shelf life. Vegetables contain naturally occurring food preservatives such as benzoic and propionic acids. Comprehensive research has been conducted on 939 vegetables from brassica, leafy vegetables, stem vegetables, and root vegetables. Most of the vegetables contained only one of the compounds. Propionic acid was found in more samples (304 samples) than benzoic acid (102 samples). As a result, perilla leaves and ginseng had the highest amount of benzoic and propionic acids, respectively [47]. Aside from these acids, vegetables have antioxidant and antimicrobial activities due to their bioactive compounds. Among vegetables, onion and garlic have widespread use in daily meals around the world as the main ingredients of most dishes. Along with their well-known antimicrobial properties, antioxidant and cytotoxic effects were also determined from their extracts [48]. Beets have been extensively used in manufactured foods recently as a colorant as a substitute for synthetic dyes and red color from cochineal bugs. Its antioxidant activity and antimicrobial effect on major pathogens were also detected for the purpose of use in food products [49, 50]. Aside from those root vegetables, the leafy vegetable group, also known as *Brassica*, has shown antimicrobial and antioxidant activity [51]. The importance of this group of vegetables is for their common usage in salads with other raw products, which makes them more susceptible to contamination with foodborne pathogens.

Whole grain cereals such as wheat, oat, barley, and millet have been added to food products as a means of a healthy diet. They are also used in fermented foods, and upon fermentation bioactive compounds may become more nutritionally available. They contain majorly phenolic acids, flavonoids, and tannins as phenolic compounds which have antioxidant effects [52]. In legumes, antioxidant activity is originated from phenolics, saponins, and peptides [53]. On the other hand, in oilseeds and edible oils, tocols, sterols, and phenolic compounds are the reason for their antioxidant activity. Mastic gum is used as a flavoring in especially bakery products, coffee, and desserts. Its substantial antioxidant activity and antimicrobial activity against major pathogens were determined in the essential oil of mastic gum [54]. The plant leaves of *Camellia sinensis* are consumed at different fermentation levels as green, oolong, and black tea. As a beverage example of the preservation effect, the antioxidant activity of this plant is due to its phenolic compounds; moreover, catechin content plays a role as an antimicrobial against several pathogens such as *L. monocytogenes* [55].

9.4 Conclusion

For a long time, various food ingredients such as spices and herbs have been used as natural food preservatives in order to increase the shelf life of the products. For this purpose, the spices or herbs have been used directly as an ingredient or their extract, essential oil, or isolated compounds with preservative effects have been added directly to food products, packaging, or coating material. The extracts and essential oils of spices and herbs show antioxidant and antimicrobial activity due to their phytochemicals such as polyphenols, flavonoids, tannins, alkaloids, terpenoids, lectins, and polypeptides. For this reason, the application of herb extract or essential oil for coating purposes or directly into the food has greatly contributed to enhancing the shelf life of food items by preventing microbial growth and oxidative reactions. Post-harvest microbial deterioration of raw materials such as tomato, mushroom, spinach, and broccoli can be decreased by the biobased film application which includes cumin, fennel, and rosemary essential oils. Adding herbs such as mint, red pepper, caraway, and cinnamon or their essential oil to the yogurt, cheese, buttermilk, milk, and so on increased the microbiological safety and shelf life of the products. In view of such desirable antimicrobial and antioxidant activity of herbs and spices against microorganisms and oxidative reactions in the meats, bread, and biscuit, they are also effective in the decreasing of biogenic amines during the production of sausages and acrylamide formation in bread and biscuits. To utilize their preservation properties, fruits and fruit-based products find their applications in sweet products such as protein bars, beverages, and bakery products. Whole grain cereals such as wheat, oat, barley, and millet contain majorly phenolic acids, flavonoids, and tannins as phenolic compounds which have an antioxidant effect. As a result, fruits, vegetables, cereals, spices, herbs, and their parts (leaves, stems, flowers, fruits, roots, barks, etc.) and wastes show antioxidant and/or antimicrobial activity due to their valuable components such as phytochemicals. For this reason, the use of natural preservatives is important in terms of both reducing the side effects of chemical preservatives and evaluating waste and excess production.

References

1. Gokoglu, N. Novel natural food preservatives and applications in seafood preservation: A review. Journal of the Science of Food and Agriculture 2019, 99(5), 2068–2077.
2. Beya, M. M.; Netzel, M. E.; Sultanbawa, Y.; Smyth, H.; Hoffman, L. C. Plant-based phenolic molecules as natural preservatives in comminuted meats: A review. Antioxidants 2021, 10(2), 263.
3. Zazharskyi, V. V.; Davydenko, P.; Kulishenko, O.; Borovik, I. V.; Brygadyrenko, V. V. Antimicrobial activity of 50 plant extracts. Biosystems Diversity 2019, 27(2).
4. Nagy, M.; Socaci, S. A.; Tofana, M.; Pop, C.; Muresan, C.; Pop, A. V.; Salanta, L.; Rotar. A. M. Determination of total phenolics, antioxidant capacity and antimicrobial activity of selected aromatic spices. Bulletin UASVM Food Science and Technology 2015, 72(1).
5. AguirrezaÂbal, M. M.; Mateo, J.; DomõÂnguez, M. C.; ZumalacaÂrregui, J. M. The effect of paprika, garlic and salt on rancidity in dry sausages. Meat Science 2000, 54, 77–81.
6. Wahba, N. M.; Ahmed, A. S.; Ebraheim, Z. Z. Antimicrobial effects of pepper, parsley, and dill and their roles in the microbiological quality enhancement of traditional Egyptian Kareish cheese. Foodborne Pathogens and Disease 2010, 7(4), 411–418.
7. Eskandari, S.; Hosseini, H.; Gholamzadeh, M.; Mousavi Khaneghah, A.; Hosseini, E. Home food safety practice and household food insecurity: A structural equation modeling approach. Journal of Food Safety 2015, 5(2), 154–160.
8. Ehsani, A.; Hashemi, M.; Naghibi, S. S.; Mohammadi, S.; Khalili Sadaghiani, S. Properties of Bunium persicum essential oil and its application in Iranian white cheese against Listeria monocytogenes

and Escherichia coli O157:H7. Journal of Food Safety 2016, 36(4), 563–570.

9. Bharti, S. K.; Pathak, V.; Alam, T.; Arya, A.; Singh, V. K.; Verma, A. K.; Rajkumar, V. Materialization of novel composite bio-based active edible film functionalized with essential oils on antimicrobial and antioxidative aspect of chicken nuggets during extended storage. Journal of Food Science 2020, 85(9), 2857–2865.

10. Ojagh, S. M.; Rezaei, M.; Razavi, S. H.; Hosseini, S. M. H. Effect of chitosan coatings enriched with cinnamon oil on the quality of refrigerated rainbow trout. Food Chemistry 2010, 120, 193–198.

11. Arancibia, M.; Gimenez, B.; Lopez-Caballero, M. E.; Gomez-Guillen, M. C.; Montero, P. Release of cinnamon essential oil from polysaccharide bilayer films and its use for microbial growth inhibition in chilled shrimps. LWT - Food Science and Technology 2014, 59, 989–995.

12. Hassanien, M. F. R.; Mahgoub, S. A.; El-Zahar, K. M. Soft cheese supplemented with black cumin oil: Impact on food borne pathogens and quality during storage. Saudi Journal of Biological Sciences 2014, 21(3), 280–288.

13. Karimirad, R.; Behnamian, M.; Dezhsetan, S. Development and characterization of nano biopolymer containing cumin oil as a new approach to enhance antioxidant properties of button mushroom. International Journal of Biological Macromolecules 2018, 113, 662–668.

14. KonukTakma, D.; Korel, F. Active packaging films as a carrier of black cumin essential oil: Development and effect on quality and shelf-life of chicken breast meat. Food Packaging and Shelf Life 2019, 19, 210–217.

15. Hemmatkhah, F.; Zeynali, F.; Almasi, H. Encapsulated cumin seed essential oil-loaded active papers: Characterization and evaluation of the effect on quality attributes of beef hamburger. Food and Bioprocess Technology 2020, 3, 533–547.

16. Caleja, C.; Barros, L.; Antonio, A. L.; Ciric, A.; Soković, M.; Oliveira, B. P. P.; Santos-Buelga, C.; Isabel Ferreira, I. C. F. R. Foeniculum vulgare Mill. as natural conservation enhancer and health promoter by incorporation in cottage cheese. Journal of Functional Foods 2015, 12, 428–438.

17. Maghami, M.; Motalebi, A. A.; Anvar, S. A. A. Influence of chitosan nanoparticles and fennel essential oils (Foeniculum vulgare) on the shelf life of Huso huso fish fillets during the storage. Food Science & Nutrition 2019, 7, 3030–3041.

18. Al-Ansi, W.; Mahdi, A. A.; Al-Maqtari, Q. A.; Fan, M.; Wang, L.; Li, Y.; Qian, H.; Zhang, H. Evaluating the role of microwave-baking and fennel (Foeniculum vulgare L.)/nigella (Nigella sativa L.) on acrylamide growth and antioxidants potential in biscuits. Journal of Food Measurement and Characterization 2019, 13, 2426–2437.

19. Naeem, A.; Abbas, T.; Ali, T. M.; Hasnain, A. Effect of antioxidant and antibacterial properties of guar gum coating containing spice extracts and its application on tomatoes (Solanum lycopersicum L.). Journal of Food Measurement and Characterization 2018, 12, 2725–2734.

20. Tiţa, O.; Constantinescu, M. A.; Tiţa, M. A.; Georgescu, C. Use of yoghurt enhanced with volatile plant oils encapsulated in sodium alginate to increase the human body's immunity in the present fight against stress. International Journal of Environmental Research and Public Health 2020, 17(20), 7588.

21. Namir, M.; Rabie, M. A.; Rabie, N. A.; Ramadan, M. F. Optimizing the addition of functional plant extracts and baking conditions to develop acrylamide-free pita bread. Journal of Food Protection 2018, 81(10), 1696–1706.

22. Da Cruz Almeida, E. T.; Barbosa, I. M.; Tavares, J. F.; Barbosa-Filho, J. M.; Magnani, M.; de Souza, E. L. Inactivation of spoilage yeasts by Mentha spicata L. and M. × villosa Huds. Essential Oils in Cashew, Guava, Mango, and Pineapple Juices. Frontiers in Microbiology 2018, 9, 1111.

23. Šojić, B.; Tomović, V.; Kocić-Tanackov, S.; Kovačević, D. B.; Putnik, P.; Mrkonjić, Z.; Đurović, S.; Jokanović, M.; Ivić, M.; Škaljac, S.; Pavlić, B. Supercritical extracts of wild thyme (Thymus serpyllum L.) by-product as natural antioxidants in ground pork patties. LWT 2020, 130, 109661.

24. Alvarez, M. V.; Ortega-Ramirez, L. A.; Silva-Espinoza, B. A.; Gonzalez-Aguilar, G. A.; Ayala-Zavala, J. F. Antimicrobial, antioxidant, and sensorial impacts of oregano and rosemary essential oils over broccoli florets. Journal of Food Processing and Preservation 2019, 43, e13889.

25. Pedrosa, M. C.; Ueda, J. M.; Melgar, B.; Dias, M. I.; Pinela, J.; Calhelha, R. C.; Ivanov, M.; Sokovi´c, M.; Heleno, S.; da Silva, A. B .; Carocho, M.; Ferreira, I. C. F. R.; Barros, L. Preservation of chocolate muffins with lemon balm, oregano, and rosemary extracts. Foods 2021, 10, 165.

26. Abedi, A.; Lakzadeh, L.; Amouheydari, M. Effect of an edible coating composed of whey protein concentrate and rosemary essential oil on the shelf life of fresh spinach. Journal of Food Processing and Preservation 2021, 45, 15284.

27. Vafania, B.; Fathi, B.; Soleimanian-Zad, S. Nanoencapsulation of thyme essential oil in chitosan-gelatin nanofibers by nozzle-less electrospinning and their application to reduce nitrite in sausages. Food and Bioproducts Processing 2019, 116, 240–248.

28. Gonçalves, N. D.; Pena, F. L.; Sartoratto, A., Derlamelina, C.; Duarte, M. C. T.; Antunes, A. E. C.; Prata, A. S. Encapsulated thyme (Thymus vulgaris) essential oil used as a natural preservative in bakery product. Food Research International 2017, 96, 154–160.

29. Lee, S.; Kim, H.; Beuchat, L. R.; Kim, Y.; Ryu, J. H. Synergistic antimicrobial activity of oregano and thyme thymol essential oils against Leuconostoccitreum in a laboratory medium and tomato juice. Food Microbiology 2020, 90, 103489.

30. Lacombe, A.; Wu, V. C. The potential of berries to serve as selective inhibitors of pathogens and promoters of beneficial microorganisms. Food Quality and Safety 2017, 1(1), 3–12.

31. Sun, X. H.; Hao, L. R.; Xie, Q. C.; Lan, W. Q.; Zhao, Y.; Pan, Y. J.; Wu, V. C. Antimicrobial effects and membrane damage mechanism of blueberry (Vaccinium corymbosum L.) extract against Vibrio parahaemolyticus. Food Control 2020, 111, 107020.

32. Zorzi, M.; Gai, F.; Medana, C.; Aigotti, R.; Morello, S.; Peiretti, P. G. Bioactive compounds and antioxidant capacity of small berries. Foods 2020, 9(5), 623.

33. Fierascu, R. C.; Temocico, G.; Fierascu, I.; Ortan, A.; Babeanu, N. E. Fragaria genus: Chemical composition and biological activities. Molecules 2020, 25(3), 498.

34. Bora, H.; Kamle, M.; Mahato, D. K.; Tiwari, P.; Kumar, P. Citrus essential oils (CEOs) and their applications in food: An overview. Plants 2020, 9(3), 357.

35. Raspo, M. A.; Vignola, M. B.; Andreatta, A. E.; Juliani, H. R. Antioxidant and antimicrobial activities of citrus essential oils from Argentina and the United States. Food Bioscience 2020, 36, 100651.

36. Singh, B.; Singh, J. P.; Kaur, A.; Singh, N. Phenolic composition, antioxidant potential and health benefits of citrus peel. Food Research International 2020, 132, 109114.

37. Pfukwa, T. M.; Fawole, O. A.; Manley, M.; Gouws, P. A.; Opara, U. L.; Mapiye, C. Food preservative capabilities of grape (Vitis vinifera) and clementine mandarin (Citrus reticulata) by-products extracts in South Africa. Sustainability 2019, 11(6), 1746.

38. Al-Saman, M. A.; Abdella, A.; Mazrou, K. E.; Tayel, A. A.; Irmak, S. Antimicrobial and antioxidant activities of different extracts of the peel of kumquat (Citrus japonica Thunb). Journal of Food Measurement and Characterization 2019, 13(4), 3221–3229.

39. Santos, T. R. J.; de Aquino Santana, L. C. L. Antimicrobial potential of exotic fruits residues. South African Journal of Botany 2019, 124, 338–344.

40. de Medeiros, I. U. D.; de Souza Aquino, J.; de Holanda Cavalcanti, N. S.; Campos, A. R. N.; de Magalhães Cordeiro, A. M. T.; da Silva Chaves, K. S. F.; Hoskin, R. T. Characterization and functionality of fibre-rich pomaces from the tropical fruit pulp industry. British Food Journal 2020, 122(3), 813–826.

41. Khoozani, A. A.; Birch, J.; Bekhit, A. E. D. A. Production, application and health effects of banana pulp and peel flour in the food industry. Journal of Food Science and Technology 2019, 56(2), 548–559.

42. Hanani, Z. N.; Yee, F. C.; Nor-Khaizura, M. A. R. Effect of pomegranate (Punica granatum L.) peel powder on the antioxidant and antimicrobial properties of fish gelatin films as active packaging. Food Hydrocolloids 2019, 89, 253–259.

43. Roda, A.; Lambri, M. Food uses of pineapple waste and by-products: A review. International Journal of Food Science & Technology 2019, 54(4), 1009–1017.

44. Asghar, M. T.; Yusof, Y. A.; Noriznan, M.; Yaacob, M. E.; Mohd Ghazali, H.; Chang, L. S.; Manaf, Y. N. Effect of processing method on vitamin profile, antioxidant properties and total phenolic content of coconut (Cocos nucifera L.) sugar syrup. International Journal of Food Science & Technology 2020, 55(7), 2762–2770.

45. Nayik, G. A.; Gull, A. Antioxidants in Fruits: Properties and Health Benefits. Springer Nature Singapore Pte Ltd., Singapore; 2020, ISBN: 978-981-15-7285-2.

46. Radulescu, C.; Buruleanu, L. C.; Nicolescu, C. M.; Olteanu, R. L.; Bumbac, M.; Holban, G. C., Simal-Gandara, J. Phytochemical profiles, antioxidant and antibacterial activities of grape (Vitis vinifera L.) seeds and skin from organic and conventional vineyards. Plants 2020 9(11), 1470.

47. Yun, S. S.; Kim, J.; Lee, S. J.; So, J. S.; Lee, M. Y.; Lee, G.; Lim, H. S.; Kim, M. Naturally occurring benzoic, sorbic, and propionic acid in vegetables. Food Additives & Contaminants: Part B 2019, 12(3), 167–174.

48. Islam, M. T.; Sarkar, C.; El-Kers, D. M.; Jamaddar, S.; Uddin, S. J.; Shilpi, J. A.; Mubarak, M. S. Natural products and their derivatives against

coronavirus: A review of the non-clinical and pre-clinical data. Phytotherapy Research 2020, 34, 2471–2492.

49. Omogbai, B. A.; Omoregie, I. A. Chemical analysis and biological activity of natural preservative from beet root (Beta vulgaris) against foodborne pathogens and spoilage organisms. African Scientist 2019, 17(2), 135–146.

50. Lourenço, S. C.; Moldão-Martins, M.; Alves, V. D. Antioxidants of natural plant origins: From sources to food industry applications. Molecules 2019, 24(22), 4132.

51. Favela-González, K. M.; Hernández-Almanza, A. Y.; De la Fuente-Salcido, N. M. The value of bioactive compounds of cruciferous vegetables (Brassica) as antimicrobials and antioxidants: A review. Journal of Food Biochemistry 2020, 44(10), e13414.

52. Adebo, O. A.; Gabriela Medina-Meza, I. Impact of fermentation on the phenolic compounds and antioxidant activity of whole cereal grains: A mini review. Molecules 2020, 25(4), 927.

53. Conti, M. V.; Guzzetti, L.; Panzeri, D.; De Giuseppe, R.; Coccetti, P.; Labra, M.; Cena, H. Bioactive compounds in legumes: Implications for sustainable nutrition and health in the elderly population. Trends in Food Science & Technology 2021, 117, 139–147.

54. Hasheminya, S. M.; Dehghannya, J. Chemical composition and evaluation of antibacterial activity of fennel (Foeniculum vulgare Mill) seed essential oil against some pathogenic bacterial strains. Food Bioscience 2020, 34, 100510.

55. Rahardiyan, D. Antibacterial potential of catechin of tea (Camellia sinensis) and its applications. Food Research 2019, 3(1), 1–6.

56. Benmaghnia, S.; Meddah, B.; Tir-Touil, A.; Hernández, J. A. G. Phytochemical analysis, antioxidant and antimicrobial activities of three samples of dried figs (ficuscarica l.) from the region of mascara. Journal of Microbiology Biotechnology and Food Sciences 2019, 9(2), 208–215.

57. Sun, X. H.; Zhou, T. T.; Wei, C. H.; Lan, W. Q.; Zhao, Y.; Pan, Y. J.; Wu, V. C. Antibacterial effect and mechanism of anthocyanin rich Chinese wild blueberry extract on various foodborne pathogens. Food Control 2018, 94, 155–161.

58. Mukherjee, S.; Nath, S.; Chowdhury, S.; Chatterjee, P. Antimicrobial activity of garlic (Allium sativum) and its potential use in fish preservation and disease prevention. International Journal of Microbiology Research 2020, 12(7), 1879–1883.

59. Skenderidis, P.; Mitsagga, C.; Giavasis, I.; Petrotos, K.; Lampakis, D.; Leontopoulos, S.; Hadjichristodoulou, C.; Tsakalof, A. The in vitro antimicrobial activity assessment of ultrasound assisted Lyciumbarbarum fruit extracts and pomegranate fruit peels. Food Measure 2019, 13, 2017–2031.

60. Alexandre, E. M. C.; Silva, S.; Santos, S. A. O.; Silvestre, A. J. D.; Duarte, M. F.; Saraiva, J. A.; Pintado, M. Antimicrobial activity of pomegranate peel extracts performed by high pressure and enzymatic assisted extraction. Food Research International 2019, 115, 67–176.

Chapter 10

ε-Polylysine as Natural Food Preservative

Azza Mutwakil Khalid Omer
Food Safety & Biotechnology Department,
National Food Research Center, Khartoum North, Sudan
azzamutwakil@yahoo.com

ε-Polylysine is typically produced as a homopolypeptide of approximately 25–30 L-lysine residues. This compound was discovered nearly 30 years ago in Japan and is naturally present, soluble in water, biodegradable, stable at high temperatures, edible, and nontoxic, and thus it is of great importance to the food and drug industry. It is mainly used as a natural preservative because it is safe and harmless to humans and the environment. The observed biological activity of ε-polylysine has attracted a great deal of attention as a potential food preservative. ε-Polylysine shows a wide range of antimicrobial activities that improve food safety, quality, and shelf life because food degradation, or loss of safety of food and beverage products is the most important problem due to its major and long-term economic, social, and even political impact.

Natural Food Preservatives
Edited by Sourish Bhattacharya and Heba Hassan Abd-El Azim Salama

ISBN 978-981-4968-64-5 (Hardcover), 978-1-003-36776-5 (eBook)
www.jennystanford.com

ε-Polylysine is manufactured employing an aerobic fermentation process using a nontoxic microorganism *Streptomyces albulus* used as a preservative in the food industry.

10.1 Introduction

Most of the food products are perishable in nature and require protection from spoilage during their preparation, storage, and distribution to give them the desired shelf life. Because food products are now sold more often in areas far from their production sites, the need for a safe extended shelf life for these products has also expanded. Preserving food is an ongoing battle against the microorganisms that spoil the food or make it unsafe. Various food preservation systems such as heating, cooling, and adding antimicrobial compounds can be used to reduce the risk of food poisoning outbreaks; furthermore, since the late 1980s, ε-polylysine has been approved by the Japanese Ministry of Health, Labor, and Welfare as a food preservative. In January 2004, ε-polylysine became Generally Recognized as Safe (GRAS) certified in the United States [1]. ε-Polylysine is used commercially as a food preservative According to research, ε-polylysine is absorbed electrostatically on the surface of the bacteria cell, followed by stripping the outer membrane. This ultimately leads to the abnormal distribution of the cytoplasm which causes bacterial cell damage. The production of polylysine by natural fermentation was first described by researchers Shoji Shima and Heiichi Sakai in 1977. ε-Polylysine is used commercially as a food preservative in Japan and Korea and imported goods sold in the United States. Food products containing polylysine are mainly found in Japan. ε-Polylysine is commonly used in food applications such as boiled rice, cooked vegetables, soups, pasta, and fish fillets (sushi) [2]. Literary studies have reported an antimicrobial effect of beta-polylysine against yeast, fungi, and Gram-positive and Gram-negative bacteria [3].

10.2 History of ε-Polylysine

The compound was first segregated from *S. albulus* sp. Lysinopolymerus strain 346 [4–6] and is right now delivered mechanically by fermentation utilizing a frisked strain of *S. albulus* sp. Lysinopolymerus strain 346 and has been broadly used as a food additives substance [7, 8]. Since the revelation of ε-PL by [9], numerous analysts have noticed the antimicrobial action of ε-PL, and the use of ε-PL as a food added substance has been accounted for [7]. Besides, based on its antimicrobial action, there are a varieties of ε-PL molecules based on their structure and molecular weight which have various biomedical applications. The exceptional physicochemical properties of ε-PL as a polycationic compound will be appropriate to different fields other than the food processing. For a copious supply of ε-PL in the future, the current creation framework may be wasteful. Another creation interaction ought to be planned. In any case, the biosynthetic component of ε-PL has not been clarified at this point. After the disclosure of the enzymes engaged with ε-PL biosynthesis and their genes, a more proficient ε-PL creation framework may be created.

10.3 Organisms Producing ε-Polylysine

ε-PL is found as an extracellular material made by the filamentous bacterium similarly to other bacterial strains or eukaryotes, for example, *S. albulus* ssp. Lysinopolymerus strain 346, *S. albulus* strain 410, ergot animals [10, 11]. ε-PL has a wide antimicrobial action against Gram-positive and Gram-negative bacteria, yeasts, and molds [12].

10.4 ε-Polylysine Production

ε-PL is currently modernly categorized in Japan as a food added substance by a fermentation utilizing *S. albulus* and its excretes produce ε-PL in a culture medium at around pH 4.0, while ε-PL is quickly debased in a pH scope of 5.0–8.0 [13–15]. The presence of an ε-PL-debasing chemical is disadvantageous for modern ε-PL creation by *S. albulus*. Therefore, an ε-PL-corrupting protein of *S. albulus* was purged and described [16]. In contrast, ε-PL-tolerant microorganisms have been separated from soils and the ε-PL-debasing proteins from these microscopic organisms have been identified [17, 18].

10.5 Antimicrobial Effects and Mechanisms

The antimicrobial activity of ε-polylysine has been reported as being effective against a wide spectrum of microorganisms, including yeasts, fungi, Gram-positive and Gram-negative bacteria, and bacteriophages [19, 10]. Antimicrobial activity was diminished by the chemical modification of the amino groups [19]. The proposed inhibition mechanism is its electrostatic adsorption onto the cell surface of microorganisms, and based on its cationic property, eventually stripping the outer membrane and causing the abnormal distribution of cytoplasm [19]. The antibacterial effect of various L-lysine oligomesa indicated that ε-polylysine with more than 10 L-lysine remainders clearly had antibacterial activities proportional to the number of remainders or residues.

10.6 Application of ε-Polylysine

ε-PL and its subsidiaries have an assortment of utilizations in numerous fields, for example, food industries, medication, and drugs [20, 21]. In any case, ε-PL has poor cell antioxidant property, which obstructs its mechanical applications. For instance, during food processing what's more, preservation measures, staples (for example meat furthermore, fish) are perishable and exposure to change the flavor, texture, and color of food [22]. In this manner, on the off chance that ε-PL can be given the acceptable cell antioxidant action without changing the antibacterial activity, it can adequately prevent the decay and extend the shelf life, particularly for meat and its products.

10.7 Food Preservative

The antimicrobial property of ε-PL and its application as a food additive has been acknowledged worldwide since its detection. ε-PL is utilized as a food additive in numerous nations. There is developing interest in forestalling the decay of food just as in extending shelf life of usability alongside the requirement of item liability law and the presentation of a by date. Different ways to deal with sensitize meat or poultry subsequent to collecting incorporate cold and high temp water, steam pasteurization or steam vacuum treatment, managing, chemical rinses, and natural corrosive rinses with or without surfactants. Freezing and low temperature protection of food has particularly improved the quality of food. There is a developing awareness that the use of chemical preservatives may present serious health problems. Thus there is an expanding interest for utilization of natural additives. As a result of its amazing antimicrobial action and heat stabilization [7], ε-PL has a lot of consideration as a characteristic food additive. By combining ε-PL with other food added substances, the additive activity is amazingly improved [7].

The utilization of ε-PL along with food added substances, for example, glycine, vinegar, ethanol, and thiamine laurylsulfonate is viable for the safeguarding of different kinds of food. At the point when the impact of ε-PL and glycine on the safeguarding of consolidated milk was analyzed, a synergistic impact on microbial development was noticed. The synergistic impact brings about a diminishing in the aggregate sum of additives to be added to the food. In Japan, we generally see an assortment of food items containing ε-PL, particularly boiled rice in lunch boxes bought from general stores and pharmacies.

10.8 Emulsifying Agent

ε-PL can be utilized as a characteristic antimicrobial specialist; however, it tends to collaborate with proteins and acidic polysaccharides, prompting to enable the loss of its antimicrobial activity, which restricts its utilization. Additionally, ε-PL has poor emulsifying properties; along these lines, its use has been limited to starch-based food sources [23, 3]. As of late, the emulsifying action of ε-PL was improved by forming with dextran through the Maillard reaction [24]. The emulsifying capacity of formed ε-polylysine was superior to those of emulsifiers Sunsoft SE11 and Q-18S, a sucrose-unsaturated fat ester and polyglycerine ester. The emulsifying activity of the form was high in neutral pH and was not influenced even in presence of high salt concentration (1.0 M NaCl). The first antimicrobial action was additionally held in the form with dextran. In this manner, the PL-dextran form could be utilized as a bi-utilitarian added substance playing out the double part of added substance and emulsifier in industries of food.

10.9 Dietary Agent

Diabetes, hypertension and atherosclerosis, and other serious health problems are associated with obesity [25]. Obesity can be forestalled either by the admission of low fat eating routine or ingestion of a characteristic item that specifically restricts intestinal retention of dietary fat. Studies uncover that pancreatic lipase assumes a significant part in lipid assimilation from the digestive system [26]; subsequently, it is proposed that characteristic items that restrain the action of pancreatic lipase may stifle dietary fat ingestion from the small digestive system [27].

10.10 ε-Polylysine Safety and Approval

ε-PL is not harmful to people [28]. Excessive oral poisonousness in rodents and bacterial inversion being examined with traces uncovered the potential of ε-PL for its consideration as GRAS [29]. In a two-age generation study utilizing rats, the nontoxic measurement level of ε-PL was closed to 10,000 ppm. No harmfulness for propagation, neurological capacity, undeveloped and fetal turn of events, and development was accounted for with ε-PL at 30,000 ppm [29]. The pharmacokinetics of ε-PL in vivo was additionally examined [2]. Assimilation, appropriation, digestion, and discharge (ADME) examination using ^{14}C-radiolabeled ε-PL uncovered that ε-PL was ineffectively invested in the gastrointestinal lot and the greater part of the dosed radioactivity was wiped out by discharge inside autoradiography. In view of these outcomes showing its security, ε-PL was supported by the Japanese Service of Wellbeing, Work, and Government assistance as an additive in food in the last part of the 1980s. From there on, ε-PL was recorded in the Korea Food Added substances Code and has been utilized in Korea. Since long ε-PL is securely utilized in Japan as an additive for various food varieties including conventional dishes like rice and noodles [7]. In food frameworks for the maximum part of usage of ε-PL contains splashing or plunging cut fish and fish sushi at levels of 1000–5000 ppm, safeguarding of bubbled rice, noodles, soup stock, and cooked vegetable at the level of 10–500 ppm [7, 23]. ε-PL has additionally been utilized in potato salad, steamed cakes, and custard cream [2]. As of late, the US Food and Medication Organization has given ε-PL GRAS status (for the most part perceived as protected). FDA supported the utilization of ε-PL as an antimicrobial specialist in cooked or sushi rice at levels up to 50 mg/kg of rice [31]. For food protection, ε-PL can be utilized alone or in mix with other food added substances such as glycine, vinegar, ethanol, and thiamine lauryl sulfonate [7, 32]. The mix with different added substances enormously improved the conservation viability of ε-PL.

10.11 Globally Utilization

ε-Polylysine is an approved food preservative in Japan [7]. Common uses of ε-polylysine include spraying or dipping sliced fish or fish sushi at levels of 1000–5000 ppm and preservation of boiled rice, noodles, cooked vegetable, and soups at levels of 10–500 ppm [7, 23]. ε-Polylysine has also been used in potato salad, steamed cakes, and custard cream [2]. Also, it is given GRAS status by [31]. FDA approved the use of ε-polylysine in cooked or sushi rice at levels up to 50 mg/kg.

10.12 A Review of Present and Future Uses

According to the Economic Research Service of the United States Department of Agriculture, the estimated annual cost of foodborne illness caused by *Escherichia coli* O157:H7, *Listeria monocytogenes*, and *Salmonella typhimurium* was $0.9, $2.3, and $2.9 billion, respectively [30]. Regulatory agencies have published guidelines to help reduce foodborne illnesses, but outbreaks continue to occur. Therefore, continuous investigations of new antimicrobial compounds or novel strategies are needed. To date, a few studies on the effectiveness of ε-polylysine against foodborne pathogens in food systems have been reported. The present study was conducted to determine the efficacy of ε-polylysine in controlling *E. coli* O157:H7,

S. typhimurium, and *L. monocytogenes* in laboratory media as well as in a complex food system. Ready-to-eat (RTE) roast beef was chosen due to its high value, its high risk of post processing contamination due to human handling, and its frequent association with foodborne illness outbreaks of *E. coli*, *S. typhimurium*, and *L. monocytogenes*.

10.13 Conclusion

This chapter has demonstrated the broad properties and uses of polylysine as a natural food preservative in the food industry. Given the wide application display, it is assumed that the refinement of this characteristic polymer is efficient, substantial, and tradable for all current food additives.

References

1. GRAS Notice No. GRN 000135 Archived 2008-05-11 at the Wayback Machine.
2. Hiraki, J., et al. 2003. Use of ADME studies to confirm the safety of ε-polylysine as a preservative in food. Regulatory Toxicology and Pharmacology 37(2): 328–340. doi:10.1016/S0273-2300(03)00029-1. PMID 12726761.
3. Hiraki, J. 1995. Basic and applied studies on ε-polylysine. Journal of Antibacterial Antifungal Agents 23: 349–354.
4. Shima, S. and Sakai, H. 1977. Polylysine produced by Streptomyces. Agricultural and Biological Chemistry 41: 1807–1809.
5. Shima, S. and Sakai, H. 1981. Poly-L-lysine produced by Streptomyces. Part II. Taxonomy and fermentation studies. Agricultural and Biological Chemistry 45: 2497–2502.
6. Shima, S. and Sakai, H. 1981. Poly-L-lysine produced by Streptomyces. Part III. Chemical studies. Agricultural and Biological Chemistry 45: 2503–2508.
7. Hiraki, J. 2000. ε-Polylysine, its development and utilization. Fine Chemicals 29: 18–25.
8. Hiraki, J. and Suzuki, E. 1999. Process for producing e-poly-L-lysine with immobilized Streptomyces albulus. US Patent No. 5,900,363.
9. Shima, S. and Sakai, H. 1977. Polylysine produced by Streptomyces. Agricultural and Biological Chemistry 41: 1807–1809.

10. Yoshida, T. and Nagasawa, T. (2003). e-Poly-L-lysine: Microbial production, biodegradation and application potential. Applied Microbiology and Biotechnology 62: 21–26.

11. Shukla, S. C., Singh, A., Pandey, A. K., and Mishra, A. 2012. Review on production and medical applications of ε-polylysine. Biochemical Engineering Journal 65: 70–81.

12. Ye, R., Xu, H., Wan, C. et al. 2013. Antibacterial activity and mechanism of action of ε-poly-l-lysine. Biochemical and Biophysical Research Communications 439: 148–153.

13. Shima, S., Oshima, S., and Sakai, H. 1983. Biosynthesis of ε-poly-l-lysine by washed mycelium of Streptomyces albulus no. 346. Nippon Nogeikagaku Kaishi 57: 221–226.

14. Hiraki, J., Hatakeyama, M., Morita, H., and Izumi, Y. 1998. Improved epoly-l-lysine production of an S-(2-aminoethyl)-l-cysteine resistant mutant of Streptomyces albulus. Seibutsu Kogaku Kaishi 76: 487–493.

15. Kahar, P., Iwata, T., Hiraki, J., Park, Y. E., and Okabe, M. 2001. Enhancement of ε-polylysine production by Streptomyces albulus strain 410 using pH control. Journal of Bioscience and Bioengineering 91: 190–194.

16. Kito, M., Takimoto, R., Yoshida, T., and Nagasawa, T. 2002. Purification and characterization of ε-poly-l-lysine-degrading enzyme from an ε-poly-l-lysine-producing strain Streptomyces albulus. Archives of Microbiology 178: 325–330.

17. Kito, M., Onji, Y., Yoshida, T., and Nagasawa, T. 2002. Occurrence of epoly-l-lysine-degrading enzyme in e-poly-l-lysine tolerant Sphingobacterium multivorum OJ10, purification and characterization. FEMS Microbiology Letters 207: 147–151.

18. Takimoto, R., Kito, M., Yoshida, T., Onji, Y., and Nagasawa, T. 2002. Characterization of microbial enzymes catalyzing the degradation of ε-poly-l-lysine. Annu Meet Soc Biosci Biotechnol Biochem Jpn 2002:297.

19. Shima, S., Matsuoka, H., Iwamoto, T., and Sakai, H. 1984. Antimicrobial action of ε-poly-L-lysine. Journal of Antibiotics 37: 1449–1455.

20. Cai, L., Li, D., Dong, Z., Cao, A., Lin, H., and Li, J. 2016. Change regularity of the characteristics of Maillard reaction products derived from xylose and Chinese shrimp waste hydrolysates. LWT - Food Science and Technology 65: 908–916.

21. Shi, C., He, Y., Feng, X., and Fu, D. (2015). ε-Polylysine and next-generation dendrigraft poly-L-lysine: chemistry, activity, and

applications in biopharmaceuticals. Journal of Biomaterials Science, Polymer Edition, 26(18): 1343–1356.

22. Karwowska, M., Wójciak, K. M., and Dolatowski, Z. J. 2014. Comparative studies on lipid oxidation of organic model sausage without nitrite produced with the addition of native or autoclaved mustard seed and acid whey. International Journal of Food Science & Technology 49: 2563–2570.
23. Otsuka, N., Kuwahara, Y., and Manabe, K. 1992. Effect of ε-poly-lysine on preservation of boiled noodles. Journal of Japanese Society of Food Science and Technology 39: 344–347.
24. Ho, Y. T., Ishizaki, S., and Tanaka, M. 2000. Improving emulsifying ability of ε-poly-lysine by conjugation with dextran through the Maillard reaction. Food Chemistry 68(4): 449–455.
25. Hill, J. O., Melansonand, E. L., and Wyatt, H. 2000. Dietary fat intake and regulation of energy balance: Implication for obesity. Journal of Nutrition 130: 284–288.
26. Duan, R. D. 2000. Enzymatic aspects of fat digestion in the gastrointestinal tract. In Christopher, A. B., Vriese, S. (Eds). Fat Digestion and Absorption, pp. 25–46. Champaign, IL: AOCS Press.
27. Kido, Y., Hiramoto, S., Murao, M., Horio, Y., Miyazaki, T., Kodama, T., and Nakabou, Y. 2003. ε-Polylysine inhibits pancreatic lipase and suppresses postprandial hypertriacylglyceridemia in rats. Journal of Nutrition 133: 1887–1891.
28. Zinoviadou, K. G., Koutsoumanis, K. P., and Biliaderis, C. G. 2010. Physical and thermo-mechanical properties of whey protein isolate films containing antimicrobials, and their effect against spoilage flora of fresh beef. Food Hydrocolloids 24: 49–59.
29. Neda, K., Sakurai, T., Stakahashi, M., Shiychi, M., and Ohgushi, M. 1999. Two generation reproduction study with teratology test of ε-poly ε-lysine by dietary administration in rats. Japanese Pharmacology and Therapeutics 27: 1139–1159.
30. US Food and Drug Administration (USFDA). 2004. GRAS Notice 000135: ε-Polylysine. Office of Food Additive Safety.

Index